HELICOPTERS

Other Titles in ABC-CLIO's
WEAPONS AND WARFARE SERIES

HELICOPTERS

AN ILLUSTRATED HISTORY OF THEIR IMPACT

Stanley S. McGowen

A B C ✹ C L I O

Santa Barbara, California   Denver, Colorado   Oxford, England

Library of Congress Cataloging-in-Publication Data
McGowen, Stanley S., 1947–
Helicopters : an illustrated history of their impact / Stanley S. McGowen.
p. cm. — (Weapons and warfare series)
Includes bibliographical references and index.
ISBN 1-85109-468-7 (hardback : alk. paper) — ISBN 1-85109-473-3 (ebook)
1. Military helicopters. I. Title. II. Series.
UG1230.M345 2005
358.4'183—dc22
2005003243

05 06 07 08 / 10 9 8 7 6 5 4 3 2 1

This book is also available on the World Wide Web as an eBook.
Visit abc-clio.com for details.

ABC-CLIO, Inc.
130 Cremona Drive, P.O. Box 1911
Santa Barbara, California 93116-1911

This book is printed on acid-free paper.
Manufactured in the United States of America

CONTENTS

PREFACE

DEFINITION: A helicopter, also called a rotary-wing aircraft, may be defined as any flying machine that utilizes rotating wings to provide lift, propulsion, and control forces that enable the aircraft to rise vertically and hover above the ground. A helicopter flies by means of thrust (lift) generated by a set of wings (blades) rotating around a shaft protruding above the fuselage of the aircraft. The spinning blades create an airflow over them, which produces lift. Pilots maneuver the helicopter by varying the pitch, or angle, of the rotor blades as they move through the air. Reciprocating (piston) and gas turbine engines provide the power on modern helicopters to turn the blades through a transmission and connecting shafts. A true helicopter can lift off vertically, fly forward, backward, or sideways, climb to altitude, cruise at speed, and then descend to a hover before landing—feats achieved in nature only by the hummingbird or the dragonfly.

Many aviation experts consider the helicopter the most innovative and versatile vehicle known to man. Although practical helicopters first flew just over sixty years ago, they are without question one of the most vital aircraft in the world today. Various types of helicopters manufactured by numerous companies, located in several nations, operate in the most remote areas of the world, inaccessible to other aircraft or vehicles. From the loftiest mountain peaks, to the most inhospitable jungles, over the seas, and to the frozen poles, helicopters function admirably, despite the most adverse weather and terrain.

As early as 400 B.C., Chinese children played with a toy made of two feathers joined together that, when spun rapidly, rose a few feet into the air and then gently descended to earth. Nineteen hundred years later the fertile brain of Leonardo da Vinci deduced that man could not fly by magic but by physical laws; consequently, he scientifically designed the first helicopter.

With a mixture of genius and determination, early twentieth-century visionaries translated da Vinci's drawings into reality. Although Russian emigre Igor I. Sikorsky is, deservedly, given much credit for developing the helicopter, the aircraft, like the fixed-wing airplane, had no one particular inventor. Frenchmen Paul Cornu and Etienne Oehmichen; Spaniard Juan de la Cierva in autogyros; German Doktor Heinrich Karl Johann Focke, in autogyros and helicopters; Soviet Nicolai I. Kamov in both types of aircraft; and Americans Arthur M. Young and Frank N. Piasecki all made significant contributions to the development of the modern helicopter. In the last half-century, from rickety, unstable machines that would hardly lift their own weight, modern aeronautical engineers have increased helicopter speeds to more than 350 mph and lifting capacity (payload) to over 60,000 pounds.

Introduced into combat during World War II, helicopters were used extensively for the first time in the Korean War, where their invaluable contributions caused many generals and admirals to re-evaluate the impact of this unconventional aircraft. Combat proven by the United States in Vietnam, the helicopter proved a crucial asset to the French in Algeria, the Israelis in several conflicts, the British in the Falklands, and several countries during Desert Storm.

Today the helicopter is a fundamental weapon on the modern battlefield. There is not a conflict in the world today that does not involve the tactical, or sometimes strategic, use of helicopters.

Contemporary sophisticated rotary-wing aircraft are employed in multifaceted military and nonmilitary roles, including air assaults; gunships (attack, antitank, and escort); antiship and submarine; special warfare operations; electronic warfare; air ambulance; aerial reconnaissance and observation; search and rescue; petroleum exploration and production; construction and logging cranes; fire suppression; mail service; police protection and traffic control; crop dusting and fertilizer spreading; cattle control and fence mending; and any other use an ingenious owner, pilot, or business operator may concoct.

ACKNOWLEDGMENTS

To ACHIEVE ANY MODICUM OF SUCCESS in a work of this magnitude, an author must rely on a number of military, private, and corporate contributors. As a blind author, I must first express my gratitude to my readers and research assistants, Jennie Tucker, Nancy Grace, and especially my wife, Jolene.

A number of military and civilian aviation authorities who contributed to bringing this project to fruition receive my sincerest appreciation. Lieutenant General Robert R. Williams, U.S. Army Ret.; Lieutenant Commander Dennis Dickson, USCG; Captain Gary Frazier, British Army Air Corps; Hans Weichsel, retired vice president, Bell Helicopter; Michael H. Cox, manager of media relations, Bell Helicopter/Textron; Garry Bass, director, Army Aviation Business Development, FLIR Systems, Inc.; Jeffery Scott Adair, Boeing Helicopters; Daniel Libertino, and the staff of the Sikorsky Archives; Lieutenant Colonel Glenn Carr, U.S. Army Ret.; CW5 Kenneth Donahue, U.S. Army Ret.; and CW3 Jonathan D. Case, USAR, all gave willingly of their time and expertise. No matter how busy with their own jobs and pursuits, they all responded quickly to my queries for the most minuscule information.

Lastly I must thank Alicia Merritt and the staff of ABC-CLIO for their forbearance during the research, writing, and publication of this encyclopedia.

Researching and writing this volume has expanded my knowledge of helicopters beyond my wildest dreams. As a helicopter pilot myself, I thought I had a fundamental knowledge of the history of rotorcraft, but I had much to learn.

This much I do know:

Standing around on the ground waiting to die is no way to live.

Stanley S. McGowen
2004

INTRODUCTION TO
ENCYCLOPEDIAS OF WEAPONS
AND WARFARE SERIES

WEAPONS BOTH FASCINATE AND REPEL. They are used to kill and maim individuals and to destroy states and societies, and occasionally whole civilizations, and with these the greatest of man's cultural and artistic accomplishments. Throughout history tools of war have been the instruments of conquest, invasion, and enslavement, but they have also been used to check evil and to maintain peace.

Weapons have evolved over time to become both more lethal and more complex. For the greater part of man's existence, combat was fought at the length of an arm or at such short range as to represent no real difference; battle was fought within line of sight and seldom lasted more than the hours of daylight of a single day. Thus individual weapons that began with the rock and the club proceeded through the sling and boomerang, bow and arrow, sword and axe, to gunpowder weapons of the rifle and machine gun of the late nineteenth century. Study of the evolution of these weapons tells us much about human ingenuity, the technology of the time, and the societies that produced them. The greater part of technological development of weaponry has taken part in the last two centuries, especially the twentieth century. In this process, plowshares have been beaten into swords; the tank, for example, evolved from the agricultural caterpillar tractor. Occasionally, the process is reversed and military technology has impacted society in a positive way. Thus modern civilian medicine has greatly benefitted from advances to save soldiers' lives, and weapons technology has impacted such areas as civilian transportation or atomic power.

Weapons can have a profound impact on society. Gunpowder weapons, for example, were an important factor in ending the era of the armed knight and the Feudal Age. They installed a kind of rough democracy on the battlefield, making "all men alike tall." We can only wonder what effect weapons of mass destruction (WMD) might have on our own time and civilization.

This series will trace the evolution of a variety of key weapons systems, describe the major changes that occurred in each, and illustrate and identify the key types. Each volume begins with a description of the particular weapons system and traces its evolution, while discussing its historical, social, and political contexts. This is followed by a heavily illustrated section that is arranged more or less along chronological lines that provides more precise information on at least 80 key variants of that particular weapons system. Each volume contains a glossary of terms, a bibliography of leading books on that particular subject, and an index.

Individual volumes in the series, each written by a specialist in that particular area of expertise, are as follows:

Ancient Weapons
Medieval Weapons
Pistols
Rifles
Machine Guns
Artillery
Tanks
Battleships
Cruisers and Battle Cruisers
Aircraft Carriers
Submarines
Military Aircraft, Origins to 1918
Military Aircraft, 1919–1945
Military Aircraft in the Jet Age
Helicopters
Ballistic Missiles
Air Defense
Destroyers

We hope that this series will be of wide interest to specialists, researchers, and even general readers.

Spencer C. Tucker
Series Editor

HELICOPTERS

CHAPTER 1

Early Developments, 1900–1939

HISTORICAL BACKGROUND

Leonardo da Vinci, the distinguished Italian artist, mathematician, and scientist, observed that birds control their flight by manipulating the tips of their wings. In 1483, with this concept in mind, he sketched a flying machine based on the Archimedes screw. Da Vinci described his wire-framed craft as an "instrument made with a helix . . . of flaxen linen of which one has closed the pores with starch, and is turned with a great speed, the said helix is able to make a screw in the air and to climb high" (Gregory 1944, 8–10). In his description da Vinci used the Greek word *helix*, meaning spiral or twist, applying the word to the theory of flight. Others followed his lead and used the term in describing their flying machines. Da Vinci also described a fabric-covered frame for his flying machine, a construction method utilized centuries later by both airplane and helicopter inventors (Gregory 1944, 6–7).

Intrigued by nature and da Vinci's designs, numerous inventors produced models that worked theoretically but would have been impractical in full-size form. In approximately 1754, Russian inventor Mikhail Lomonosov fashioned a small coaxial rotor modeled after the Chinese toy but powered by wound-up springs. When released, his contrivance flew for a few seconds. In 1783 the French naturalist Launoy, assisted by his mechanic, used turkey feathers to construct a coaxial version of the Chinese top. When their model fluttered into the air it stirred considerable interest among certain scholars.

Sir George Cayley, known for his work on the basic principles of flight in the 1790s, had constructed several successful vertical-flight models by the end of the eighteenth century. He cut rotors from sheets of tin and powered his models with coiled springs. In 1804 he constructed a whirling-arm device that he used to scientifically study the aerodynamic forces produced by lifting surfaces. In 1843, Cayley published a scientific paper in which he theorized a relatively large vertical-flight aircraft that he called an "aerial carriage." His concept, however, remained only speculation, because the steam engines available at the time were far too heavy for powered flight.

Although inventors who powered their models with miniature, lightweight steam engines enjoyed some limited success, the lack of a suitable power plant stifled aeronautical progress for decades. In the mid-1800s another Englishman, Horatio Phillips, unveiled a vertical-flight machine powered by a tiny boiler. Steam generated by the diminutive engine was ejected out of the blade tips, providing thrust to turn the rotors. Certainly impractical at full-scale, Phillips's experiment was nonetheless significant because it recorded the first instance in which an engine, not stored energy devices, powered a flying model helicopter. In the early 1860s, Frenchman Ponton d'Amecourt flew several small steam-powered models that he called "helicopters." The word, derived from the Greek adjective *elikoeioas*, meaning "spiral" or "winding," and the noun *pteron*, meaning "feather" or "wing," led to the modern term "helicopter."

Many pioneers of vertical flight developed original designs, but all the early experimenters lacked two essentials: a true understanding of the nature of lift (aerodynamics) and an adequate power source. Aviation records document a myriad of failed rotary-wing machines. Most malfunctioned because of poor aerodynamic or mechanical design, or an inadequate power source; some just vibrated themselves to pieces.

Few who contemplated fabricating a helicopter realized the true complexities of vertical flight, and most failed to approach their venture in a scientific manner. In the 1880s, however, U.S. inventor Thomas Alva Edison experimented with small helicopter models, testing several rotor configurations driven by a gun cotton (that is, nitrocellulose) engine. An early form of the internal combustion engine, Edison's gun cotton engines exploded during a series of experiments; gun cotton was an ingredient for both gunpowder and dynamite. Edison switched to a less volatile electric motor for his later experiments. His more scientific approach indicated that both high lift coefficient from the rotor system and a high power source were required to sustain vertical flight.

A significant technological breakthrough came at the end of the nineteenth century with the invention of the internal combustion engine. By the 1920s, advancements in metallurgy produced lighter engines with high power-to-weight ratios. Early internal combustion engines were made of cast iron, but after World War I aluminum became more widely used in aviation applications, allowing construction of full-size helicopters with a lightweight power source. Unfortunately for inventors, increased power did not solve their problems but only exacerbated the complexities of vertical flight.

In addition to misunderstandings surrounding the basic aerodynamics of vertical flight and a suitable power plant, several other factors inhibited the production of a successful helicopter:

Torque. Torque is the effect produced by a single rotor to force the fuselage to rotate in a direction opposite from the main rotor. To counter the torque of a single main rotor, early inventors constructed machines with coaxial, counter-rotating main rotors, rotors mounted laterally from the aircraft body, a smaller tail, or an anti-torque rotor.

Dissymmetry of lift. This phenomenon, which tended to cause experimental helicopters to flip over, continually confounded the early designers and pilots. Dissymmetry of lift is the unequal lift produced by rotating helicopter blades when in forward flight. One blade advances into the relative wind, creating more lift, while the other retreats, producing less lift. The invention of the swashplate solved many of the problems associated with dissymmetry of lift. The swashplate, along with blade articulation in the form of flapping and lead/lag hinges, allowed a pilot to alter the pitch angles of the swirling rotor blades. These control measures induced higher pitch angles in the retreating blade, which caused it to produce lift equivalent to that of the advancing blade, thus providing equal lift on each side of the rotor disk. In 1906, Italian Gaetano A. Crocco recognized the need to balance aerodynamic loads between the rotors; he patented an early method of pilot cyclic pitch control.

Vibration. High vibrations generated many mechanical failures of both rotors and airframes, demonstrating an obvious lack of comprehension of the dynamics of rotating components. In many instances, only trial and error taught inventors and pilots how to rid their machines of catastrophic vibrations. By adding weight to the side of driveshafts and placing the bolts of clamps 90 degrees apart along the driveshafts, mechanics smoothed out some high-frequency vibrations. Balancing each rotor blade quieted low-frequency vibrations.

LIMITED SUCCESSES

Engineering problems with vertical lift baffled inventors until, on August 24, 1907, French inventor Louis C. Breguet momentarily flew his machine, the first helicopter to leave the ground with a pilot aboard. Breguet's ungainly aircraft consisted of a set of four slowly rotating biplane wings at each end of a cross-shaped tubular frame. The thirty-two lifting surfaces afforded adequate lift, but Breguet's contrivance lacked any real means of control. During the next two years he built two improved versions but achieved little more success with controllability. Breguet's concept, nonetheless, substantiated the possibility of lifting man and machine off the ground using rotating wings.

On November 13, 1907, another French pioneer, bicycle maker Paul Cornu, lifted a better designed twin-rotored helicopter into the air for about twenty seconds. Cornu's machine consisted of four bicycle wheels supporting a puny framework that mounted a two-bladed rotor system at either end. An Antoinette 24-horsepower airplane engine drove the rotors through a system of belts. Cornu compensated for torque with counter-rotating rotors. Auxiliary wings provided rudimentary control, and, although controllability remained questionable, his flight went on record as the first controlled vertical free flight.

Not all early experiments with vertical flight occurred in Europe. A U.S. father and son team, Emile and Henry Berliner, began testing vertical-flight models in 1909. During their experiments the pair discovered that the power required to maintain a hover was substantially greater than that required to fly at low airspeeds. The phenomenon they observed, but could not define, is termed translational lift by modern aeronautical engineers. As a helicopter moves slowly forward from a hover into forward flight, the airflow through the rotor disk changes directions. It shifts (translates) from vertical to passing over the rotor blades, much like an airplane wing, making the rotors more efficient and thus reducing the power required to keep the helicopter airborne. In 1919, Henry Berliner built and tested a full-size counter-rotating coaxial rotor machine that made brief hops

about 4 feet into the air. By the early 1920s, the Berliners successfully flew an aircraft with side-by-side rotors. They attained directional control by differential longitudinal tilt of the rotor shafts.
Wings attached under the rotor system aided in lateral control. The
Berliners utilized a conventional elevator and rudder assembly at the
tail for pitch and yaw control, augmented by a small vertically
thrusting rotor on the rear of the fuselage. Their machines never attained real success, however, and the Berliners abandoned the helicopter in favor of a hybrid machine they termed a helicoplane. Although posting no remarkable success, the Berliner's early flights
with both coaxial and side-by-side rotor machines went down as
some of the first piloted helicopters in the United States.

Following Cornu's milestone, many European inventors produced
several new and unique designs, including Austrian Stephan Pe-
troczy and the renowned Hungarian-born aerodynamicist Theodore
von Kármán, who built and flew a coaxial rotor helicopter between
1917 and 1920. Designed to replace observation balloons in the
Austro-Hungarian Army, the unusual aircraft featured a station
above the rotors for a pilot and observer, inflated bags for landing
gear, and a quick-opening parachute in the event of an emergency.
Three rotary engines powered the peculiar device. Although it never
flew freely, the apparatus did lift off on several vertical flights while
tethered to the ground by cables.

Although numerous inventors experimented with helicopters during World War I, no great advances in vertical flight occurred until
1924. During the early 1920s, Etienne Edmond Oemichen, a young
French engineer at the Peugeot Company, began experimenting
with rotary-wing aircraft. On November 11, 1922, his second machine lifted off the ground in vertical flight. This model had an X-
shaped latticework tubular frame with a wide two-bladed rotor at
the end of each arm. Oemichen added eight small variable-pitch
and reversible propellers for lateral stability and control. Five horizontal propellers provided lateral stability, one at the nose afforded
steering control, and a pair of pushers drove the aircraft forward.
Through 1923, Oemichen flew this craft several times, gradually extending his time of flight. On April 14, 1924, he established a rotary-wing distance record of 1,181 feet. On May 4 he became the
first to fly a helicopter 1 kilometer (3,280 feet) in a closed circuit.
For that seven minute and forty second flight, Oemichen won a
prize of 90,000 francs. Despite his helicopter's exhibiting sufficient
controllability and power in ground effect, it attained a forward airspeed of only 4.9 mph. Recognizing his aircraft's impracticability,

Oemichen began pursuing a series of helicopters with a single main rotor.

Around 1911, Russian inventor Boris Yuriev also attempted to build a single-rotor helicopter. Yuriev's configuration resembled modern helicopters, with a large single main rotor and a smaller rotor at the end of the tailboom. The large-diameter blades suggested that the Russian understood the need for high aerodynamic efficiency to produce sufficient lift. As with most primitive helicopters, however, Yuriev's model was too heavy and lacked an engine powerful enough to lift the airframe. One of the first to employ an antitorque tailrotor, Yuriev was also one of the first to incorporate a pilot cyclic pitch apparatus for main rotor control.

In 1919, Spanish engineer Marquis Raul Pateras Pescara undertook the construction of a functional rotary-wing aircraft. In 1922 he moved his operations from Barcelona to Paris and entered the competition for the 90,000-franc prize for a 1-kilometer helicopter flight. A clumsy affair, his first machine weighed more than 1,300 pounds without fuel or pilot. A tiny 45-horsepower Hispano engine powered the large coaxial rotor system. Each of the two rotors measured 21 feet (6.4 meters) in diameter and consisted of six pairs of biplane blades, a total of twenty-four lifting surfaces. At maximum throttle the diminutive Hispano motor could not lift the machine off the ground. In June 1922, Pescara unveiled a conceptualized coaxial rotor aircraft, in which the rotors could stop in flight, transforming the aircraft into a biplane. He built a small-scale, pusher propeller powered model of the concept, but the project halted there.

Pescara's third machine, completed in 1923, could be described as a true helicopter. The machine utilized four 23.6-foot- (7.2-meter-) diameter four-blade biplane rotors, with no other means of propulsion. In the first effective use of cyclic and collective pitch control, the pilot could alter the pitch of the sixteen blades in flight by warping, or twisting, the wings. (Wilbur and Orville Wright used wing warping to control their airplanes.) Pescara provided the pilot with a cyclic stick for forward and lateral movement, a collective lever for increased/decreased lift, and a wheel to control yaw. Cyclic, collective, and antitorque arrangements are required components of a true helicopter control system. The pilot tilted the rotor hub on Pescara's craft to move the ship into forward flight, but airspeed was less than 10 mph (13 kph). To overcome low airspeeds, previous inventors had employed auxiliary propellers on their "helicopters." Exhibiting unusual forethought, Pescara also included an autorotational capability on his helicopter in the event of engine failure.

During autorotation the relative wind passing through the rotor blades comes from the underside, the airflow velocity maintaining rotor rpm, thus providing the pilot with enough rotor thrust to bring the aircraft to a safe landing. In September 1923, Pescara narrowly missed becoming the first person to fly a rotary-wing aircraft through a 1-kilometer circuit. Unfortunately he crashed, and his machine was severely damaged. On April 18, 1924, four days after Oemichen's record flight, Pescara almost doubled Oemichen's distance record. By 1932 other inventors had bettered the records only slightly. Records for helicopter flight stood at just over half a mile, only fourteen minutes in duration, with a paltry altitude of 50 feet above the ground.

A helicopter designed by a fugitive from the Bolshevik Revolution in Russia, Dr. George de Bothezat, participated in attempts to break some of those records. A student of early helicopters, Major T. H. Bane, chief of the Engineering Division of the U.S. Army Air Service, convinced his superiors that the U.S. Army should enter the international race to produce a serviceable helicopter. In January 1921, at Bane's insistence, the Army Air Corps awarded a contract to de Bothezat and Ivan Jerome to develop a vertical-flight machine; it was the first military contract awarded in the United States to manufacture a helicopter. Their large cross-shaped airframe supported a 22-foot-diameter, six-blade rotor at the end of each of four pylons. Each pylon also held two small variable-pitch propellers intended to provide thrust and yaw control. The main rotor axes were not parallel but inclined slightly inward, to prevent the tips from intermeshing at a point directly above the center of the airframe. Another small rotor, mounted above the 180-horsepower Le Rhone radial engine bolted in at the junction of the frames, was expected to provide some lift and cooling for the engine. Deemed superfluous, the designers later removed that rotor.

The complex control system provided each main rotor with collective pitch control that produced differential thrust, thus allowing the pilot to tilt the airframe in any direction. By inclining the machine in the desired direction, the pilot could, hypothetically, direct the aircraft laterally over the ground. The contract stipulated that the aircraft be able to descend with reduced throttle, and de Bothezat included a method for autorotation, albeit very complicated. The pilot had to pull another control lever to switch into negative pitch to enter autorotation, increasing the pilot's workload. At takeoff on its first flight in October 1922, the cumbersome apparatus weighed 3,740 pounds and managed to remain aloft for one minute and forty-two seconds. To provide additional power, de Bothezat and Jerome upgraded the engine to a 220-horsepower Bentley rotary motor.

At McCook Field, known today as Wright-Patterson Air Force Base, near Dayton, Ohio, the team made about a hundred short test flights in 1923 before abandoning the project. In April the helicopter rose to a low hover with four "passengers" hanging onto the airframe. Although relatively stable at a hover, testing confirmed that de Bothezat's design was underpowered, unresponsive, and mechanically complex, resulting in demonstrated unreliability. Pilot workload was too high during hover to attempt lateral motion. Additionally, the contract specified a 300-foot hover, and the highest the craft ever managed was about 15 feet. De Bothezat's ungainly helicopter's stability did, however, establish that a practical helicopter was theoretically possible. With the poor performance of de Bothezat's aircraft, however, the Army Air Corps quickly lost interest in the project in favor of autogyros. In the late 1930s, de Bothezat tried again with a single seat, coaxial helicopter, but he achieved even less success than with his first endeavor.

Contemporaneous with the de Bothezat trials, two U.S. inventors, Professor Francis B. Crocker and Dr. Peter Cooper Hewitt, experimented with a helicopter design of their own. In early 1924 the ship, which had two large wings that slowly rotated in opposite directions, lifted off the ground for a short flight. Although crudely designed, the Cooper Hewitt machine achieved a relatively promising lift-power ratio, capable of lifting its own weight and a pilot weighing up to 250 pounds. Regardless of the initial flights the aircraft guaranteed little success, and the inventors abandoned their venture.

In Great Britain, Louis Brennan introduced a novel idea to counteract the effects of torque on a single-rotor helicopter. He attached small propellers to the ends of the blades themselves. For flight control, Brennan installed servo-flaps, or ailerons, inboard of the propellers. He initially conducted his program in secrecy, and in 1921 the device successfully lifted off inside a balloon shed. Brennan unveiled his project in 1925 and undertook several flights outdoors at low altitude. After a crash on its seventh flight, in which the machine was virtually destroyed, official interest in Brennan's machine quickly waned in favor of British interest in autogyros.

In 1930, Maitland Bleeker of the United States followed Brennan's concept by mounting propellers at the ends of each of four blades of a single-rotor system. A complex arrangement of chains and gears from a centrally mounted engine powered the rotors.

Bleeker also utilized the control method developed by the Italian Coradino d'Ascanio, attaching control tabs at the trailing edge of each blade; Bleeker called them "stabovators." He incorporated both collective and cyclic pitch into his flight control system. His helicopter wobbled into ground effect hovers on several occasions, but unmanageable vibrations and instability caused Bleeker to abandon his project in 1933.

In the early 1930s, Breguet brought improved designs to his helicopter experiments. In 1931 he introduced a small coaxial superimposed twin-rotored helicopter that achieved little more success than his earlier ventures. A larger, more powerful craft of similar configuration, however, offered better prospects. The new helicopter featured a modern looking coaxial rotor system, which consisted of four tapered blades mounted to the hubs with hinges that allowed the blades to flap and lag. Breguet also integrated an innovative swashplate assembly to control the blades cyclic pitch as they swept around the mast. He installed horizontal and vertical tails to increase stability in his new craft. In this machine Breguet took off vertically, flew forward, circled, and returned to land at the starting point. He reportedly reached an altitude of approximately 500 feet, while the aircraft exhibited acceptable control. During one flight, which lasted sixty-two minutes and covered 27 miles (44 km), Breguet reached an airspeed of 50 mph. Buoyed by his helicopter's performance he proposed a huge transoceanic helicopter, but the outbreak of World War II in Europe ended his experiments. Breguet, nonetheless, founded an air transport company that eventually became Air France.

From 1910 through the 1930s many inventors from several countries built and flew rotary-wing aircraft. Most employed coaxial rotor systems with large blades, while a few dreamed of machines with more radical designs. In Denmark, watchmaker and electrical engineer Jacob Christian Ellehammer built his first helicopter model in 1911. The next year he constructed a full-size version of his compound helicopter, with a coaxial fabric-covered rotor system and a conventional propeller for propulsion. Each of the six-bladed counter-rotating rotors measured about 25 feet in diameter and were driven through a hydraulic clutch and gearbox. Realizing the need to vary blade pitch, Ellehammer included a cyclic pitch mechanism in the control system. The pilot's seat moved fore and aft to adjust. Over a period of four years Ellehammer made several short flights (some described them as more like hops) in his machine, but when it was destroyed in a 1916 crash he abandoned helicopters until 1930, when he built a rotorcraft with two superimposed rotors turning in opposite directions. Although Ellehammer flew the new machine several times, it never achieved an altitude of more than a few feet.

In The Netherlands, Doctor A. G. Von Baumhauer introduced a somewhat different idea in helicopter design, but few realized the significance of his concept at the time. In 1924, Baumhauer unveiled a helicopter with an engine at one end of a tubular frame that drove a large two-bladed lifting rotor. At the other end of the airframe another little engine turned a single small, antitorque propeller. Von Baumhauer attempted to achieve rotor control by a cyclic-pitch mechanism and by attaching the main rotor blades to a swashplate with hinges that allowed limited flapping of the individual blades. Unfortunately, he did not connect the main and tail rotors, and directional control remained difficult at best. Although the idea of a single-rotor helicopter was patented as early as 1918, Baumhauer was the first to incorporate the principle in a full-scale helicopter. Baumhauer's success with a semistabilized vehicle spawned numerous subsequent experiments with single-rotor helicopters utilizing smaller antitorque rotors. Many historians consider Baumhauer's helicopter the first step in an era of major advances in helicopter technology.

During a 1930 exposition held in Antwerp, Belgium, many aviation experts voiced considerable interest in a radical helicopter prototype exhibited by Nicolas Florine. First flown in 1933, the novel platform had two main rotors, one placed forward and the other at the feet end of the airframe. Although the rotors rotated in the same direction, Florine compensated for torque by tilting the rotors laterally in opposite directions. Each blade of the two four-bladed rotors swept down along the trailing edge near the hub, which increased the chord line, and thus the lift of the blades. A honeycomb radiator and a small propeller, driven off the main shaft, cooled a 180-horsepower Hispano Suiza engine installed amidships. Instead of the planned wheels, four stubbed rests constituted the landing gear. On its first flight the machine achieved an altitude of about 20 feet and remained aloft for eight minutes. Although the first tandem-rotored helicopter ever flown demonstrated remarkable flight characteristics and, in October 1933 set an unofficial endurance record of nine minutes and fifty-eight seconds, it never became a successful helicopter. All of Florine's models were destroyed during World War II. Frank Nicholas Piasecki, however, revived Florine's concept and launched a successful series of tandem-rotored helicopters.

Also in 1930, Coradino d'Ascanio of Italy made several flights in a helicopter of his own design. During one of those flights he set several unofficial records by remaining aloft for eight minutes and forty-five seconds, reaching an altitude of 59 feet, and covering a distance of roughly 1.75 miles (9,589 feet). D'Ascanio's machine relied on two counter-rotating twin-bladed rotors, mounted one above the other. D'Ascanio attached the rotor blades to the hub with hinges, allowing the blades to both flap and feather. To alter blade pitch he used auxiliary wings, or tabs, attached to the trailing edges of the main blades, a concept later adopted by other designers. A pilot's cyclic stick controlled these tabs through a system of cables and pulleys, allowing lateral control of the aircraft. All the tabs moved collectively to increase the rotor thrust for vertical flight. To augment pitch, roll, and yaw control, d'Ascanio added three small propellers to the airframe. At least one of those who saw it fly described the helicopter as "quite graceful in flight," and it set modest unofficial records in altitude of 57 feet (17.4 meters), duration of flight of eight minutes and forty-five seconds, and distance of 3,589 feet (1,078 meters). Insolvable vibration and instability, however, forced d'Ascanio to discontinue his project.

AUTOGYROS AND GYROPLANES

A revolutionary advancement in machines capable of vertical flight came in 1923 with the Spanish civil engineer Juan de la Cierva's C-4 autogyro. The autogyro (gryoplane) appeared as a hybrid between a helicopter and an airplane, with a conventional propeller, airframe, and tail, but with the rotors mounted above the fuselage. De la Cierva's design consisted of a converted fixed-wing aircraft with a coaxial rotor system that he hoped would solve the problem of asymmetrical lift. Unfortunately, the rotors failed to control the rolling motion of the aircraft, and his first autogyro crashed.

"Gryoplane" is an official term designated by the Federal Aviation Administration (FAA) describing an aircraft that derives lift from a freely spinning rotary wing (rotor blades), and that receives its thrust from an engine-driven propeller. Historically this type of aircraft was known as the autogyro and the gyrocopter. The early names and variants were filed as trademarks. Early autogyros were powered by engines in a tractor (pulling) configuration and were relatively heavy. Modern gyroplanes, with a rear-mounted engine and pusher propeller, are lighter and more maneuverable.

Gyroplanes exhibit qualities of both helicopters and airplanes, thus are a type of hybrid. Gyroplanes can fly at slower airspeeds than airplanes and are less likely to stall, but they cannot hover. Helicopters draw air down through engine-powered rotor systems in order to hover. Since the rotor system on a gyroplane is powered only by the air rushing over the rotor blades, much like a windmill, no antitorque rotor is required. Air velocity turning the rotor blades is known as autorotation, and it occurs in helicopters as well. Both gyroplanes and helicopters utilize this flight characteristic to descend to the ground in the event of an engine failure.

Controversy evolved over the actual date of the first flight of de la Cierva's autogyro, but on either January 9 or January 17, 1923, the first controlled gyroplane flight in history occurred at Getafe Airdrome, near Madrid, Spain. After his initial failure de la Cierva conducted wind tunnel tests on several small models and solved most of his aerodynamic problems. His fourth machine incorporated mechanical "flapping" hinges at the blade root, which somewhat equalized the lift on each side of the rotor disk. The hinges allowed the blades to flap up or down, reacting to the varying pressures exerted on the spinning blades. Although Charles Renard had suggested the principle of flapping blades in 1904, and Louis Breguet patented the idea in 1908, de la Cierva received credit for the first successful practical application to a rotating-wing aircraft.

Notwithstanding de la Cierva's successful flight, his rudimentary designs required improvements on successive models. On his first models the engine drove the propeller only. A ground crew started the rotors turning by pulling a rope wound around the rotor shaft. Otherwise the pilot taxied around on the ground until the rotors, or "windmills" as de la Cierva called them, began spinning, then the pilot opened the throttle to gain sufficient airspeed to lift off into flight. To synchronize the hinged blades, the inventor linked the rotors together with small cables and turnbuckles near the midpoint of each blade. De la Cierva also installed a clutch on later models that allowed the pilot to power the rotors with the engine, and to disengage them for autorotation.

In later models of his autogyros, de la Cierva incorporated even more sophisticated controls to compensate for asymmetrical lift. He added a lag hinge, which allowed the blades to move freely fore and aft, further alleviating the aerodynamic forces exerted on the blades. His modifications eventually evolved into a fully articulated rotor hub. A control stick allowed the pilot to tilt the entire rotor disk to fly the aircraft, eliminating the need for ailerons, but not the elevator and rudder. In 1935 de la Cierva introduced an advanced rotor-blade pitch-change mechanism in versions of the C-30 model autogyro. De la Cierva referred to his integrated control system as "orientable direct rotor control."

Although de la Cierva's autogyros exhibited some helicopter capabilities, they could not hover. His machines, nonetheless, required minimal forward airspeed to maintain flight. An inestimable number of flights proved that his autogyros were very safe and essentially stall-proof. The autogyros could land in small, confined areas because they handled well at low speeds. Takeoffs, however, required a short runway to attain flying speed. De la Cierva's autogyro was not a true helicopter, but his contribution to rotorcraft control systems was nonetheless significant. Some aviation scholars consider the first flight of the autogyros almost as momentous as the Wright brothers' success at Kitty Hawk.

In 1924, Dr. Heinrich Focke and George Wulf formed a new aircraft firm in Germany. One of the nation's most brilliant aeronautical engineers, Focke possessed a doctorate in engineering, and Wulf was a veteran test pilot. By 1931 the autogyro had become a proven vehicle. Professor Focke, noted for forward thinking and a willingness to explore unorthodox avenues of flight, recognized the inestimable value of operating an aircraft into and out of a small area. As a result of this vision, Focke-Wulf purchased a manufacturing license from the Cierva Autogyro Company. Under that license Focke-Wulf built more than thirty Cierva C-19 and C-30 autogyros.

In 1935, under Adolph Hitler's accelerating rearmament program, the German Air Ministry issued a requirement for a utility/liaison aircraft. Focke-Wulf decided to enter the competition with an autogyro. The company built and flight tested the FW-186 autogyro, which was similar to—but more streamlined than—the Cierva autogyros. A competitor, the Fieseler *Storch* (Stork), won the competition, ending further development of the FW-186.

The autogyro concept, nonetheless, proved itself useful during the 1930s and 1940s: the U.S. Postal Department used these craft for mail delivery from the roofs of post offices for nearly ten years. Hundreds of flights carrying thousands of pieces of mail were performed by Kellett and Pitcairn gyroplanes flying in Camden, New Jersey; Philadelphia, Pennsylvania; Chicago, Illinois; New Orleans, Louisiana; Washington, D.C.; and other cities.

During the same period aircraft designers in the Soviet Union developed successful autogyros. Nicolai I. Kamov built his first rotor-winged aircraft in 1929 and, along with N. K. Skrzhinskij, continued to develop autogyros into World War II. These two Soviet pioneers created several autogyros, including an armed variant. The two-place A-7-3, equipped with two machine guns, became the only armed autogyro in the world. Seven A-7s saw limited combat action as reconnaissance aircraft through 1941.

THE FIRST SUCCESSFUL HELICOPTERS IN EUROPE

In the early 1920s, Focke concluded that all previously built helicopters were just the result of trial-and-error experimentation by inventors with little or no systematic scientific thought. To prevent random errors in his developmental venture, Focke initiated a detailed research program before any design or prototype construction began. Focke defined his design objectives to eliminate several major limitations of previous helicopters, and his conclusions are still relevant to contemporary helicopter production. He outlined a plan to study the theoretical aspects of rotary-wing flight and to develop test models to verify his concepts. Only after intense testing would construction of a full-size helicopter begin (Ross 1953, 63–64).

Focke realized that autogyro technology was the starting point for his endeavor. Unfortunately, Focke-Wulf's license from de la Cierva granted rights only to manufacturing, not to Cierva's theoretical data. However, the British company Avro produced the Mk I liaison autogyro, also under license from de la Cierva, and the British government published the technical report of Glauert and C. H. N. Lock, two of the British Aeronautical Research Committee's most preeminent engineers, who had just completed a critical study of the autogyro. The British data became the basis for Focke's helicopter engineering team.

To confirm their analytical calculations, Focke and Gerd Achgelis, another leading aircraft engineer, directed the construction of a three-bladed rotor model, powered by a 3-horsepower electric motor. They tested the model under both powered and autorotational conditions. They also evaluated flight characteristics both in and out of ground effect. When a rotor system operates in ground effect, roughly half the diameter of the rotor disk above the ground, the rotor blades become more effective. The whirling blades pull air down, creating a "cushion" of air below the helicopter, thus requiring less power to hover. Few previous experimental helicopters had produced the power to rise out of ground effect, so little was known about this aerodynamic phenomenon.

Before the construction of a full-size helicopter began, the Focke-Wulf team built and flight-tested another scale model that weighed almost 11 pounds. A 0.7-horsepower, two-cylinder gasoline engine powered the rotor system. In November 1934 the model attained an altitude of 59 feet, which was then equal to the world's record for full-scale manned helicopters (Ross 1953, 92–95).

Following the scale model's successful flights, Focke-Wulf constructed and ground-tested a full-size rotor system and gearboxes. During these tests engineers measured the rotor's lifting capacity by hanging weights below the rotors. The Germans also ascertained the reliability of the gearboxes by conducting a fifty-hour test run, and then tearing down the boxes to determine wear and fatigue. With satisfactory results, construction began on a full-size aircraft.

For his full-size helicopter Focke selected the fuselage and engine of a training biplane already produced by the company. The *FW Stieglitz* was a two-cockpit aircraft with a fuselage of welded steel covered by fabric. The helicopter retained the conventional-looking vertical tail and rudder, but Focke's engineers replaced the wings and front cockpit with two three-bladed plywood rotors mounted on tubular steel outriggers. The outriggers also included a swashplate assembly, and the blades were attached to the hub with flapping and lead/lag hinges. The counter-rotating rotors eliminated torque effects and allowed the rotor system to operate without the fuselage affecting air flow through the rotors. Separating the rotors also eliminated interference from vortices created by the rotors themselves. Inclining the rotors slightly inward also increased lateral stability of the newly designated Fa-61.

A single 160-horsepower radial engine drove both rotors through a BMW transmission and a series of complicated shafts and gearboxes. A shortened propeller remained to provide cooling air for the engine during hover and slow flight. Unlike an autogyro, the propeller provided negligible forward thrust to the aircraft.

Although the Fa-61's configuration created the appearance of an autogyro with two rotors instead of one, the aircraft was a true helicopter. A robust control system allowed the pilot to tilt the rotor disks forward and backward simultaneously, causing the aircraft to move in the desired direction. By moving the cyclic stick sideways, the pilot increased the angle of the blades on one rotor and reduced the pitch on the other, thus controlling the roll of the craft. The rudder pedals tilted the rotors in opposite directions to control yaw. In forward flight the conventional tail added both longitudinal and yaw stabilization. Unlike most previous helicopters, the Fa-61 exhibited excellent stability and was capable of hovering, ascending, and descending vertically, as well as flying forward and backward.

After several tethered tests, Ewald Rohlfs piloted the Fa-61 on its first free flight on June 26, 1936. The first flight lasted less than a minute. After the initial test flight, Focke restricted the Fa-61 to only a few more indecisive flights of short duration. He then suspended the test flights in order to make several small improvements to the prototype.

On May 10, 1937, Rohlfs flew the Fa-61 on a record-breaking flight. He reached an altitude of 1,130 feet (344 meters). After setting a new altitude record he reduced the engine power to idle, entering autorotation. Although the Fa-61's control system did not provide full collective pitch control, the helicopter's freewheeling rotors allowed pilot and machine to descend safely to the ground, demonstrating conclusively that a helicopter would not inevitably crash if its engine failed.

During the next several weeks Rohlfs broke every previous helicopter record with the Fa-61. One flight extended to one hour and twenty minutes; on another the helicopter reached a speed of 76 mph (122 kph); still another extended the record to a distance of 143 miles (230 kilometers). The Fa-61 ultimately set an altitude record of 11,243 feet (3,427 meters).

The success of the Fa-61 heightened the Luftwaffe's interest in Focke-Wulf's unusual aircraft. In September 1937, Karl Francke, chief test pilot of the Reichlin experimental center, and Germany's famed female pilot, the twenty-five-year-old Flugkapitan Hanna Reitsch, received orders to conduct several flights in Focke's prototype. In subsequent flights, Reitsch broke several of Rohlfs's previous records. With the introduction of the German Focke-Wulf Fa-61, the first practical helicopter became a reality, and vertical flight was no longer a dream.

Adolph Hitler's government immediately recognized the propaganda value of having its famous female pilot fly such an advanced aircraft in public. In February 1938, Reitsch flew the Fa-61 for fourteen consecutive nights inside the huge, enclosed Deutschlandhalle sports stadium in Berlin. Although Reitsch had less than three hours flying experience with the aircraft, she adeptly demonstrated the helicopter in a confined area before large crowds, proving the exceptional maneuverability of the Fa-61.

Focke's helicopter so impressed the Luftwaffe command that it awarded his company a contract for an enlarged version of the aircraft that could lift 1,500 pounds (680 kilograms) of cargo. The new helicopter, designated the Fa-223 *Drache* (Dragon), made its first flight in the spring of 1940. A 1,000-horsepower engine drove two triple-bladed 39-foot (12-meter) rotors. The Fa-223 measured slightly more than 80 feet (24 meters) wide and 40 feet (12 meters) long and, unlike the Fa-61's open cockpit, sported a four-seat enclosed passenger cabin. Eight additional seats could be installed outside to carry passengers at low speeds. The company built seventeen prototypes and one production version before the Allies bombed the assembly line, and Focke fell out of favor with the Nazi government, effectively ending Focke-Wulf's helicopter production.

The Fa-223's performance records included a top speed of 115 mph (185 kph) and an altitude of 23,400 feet (7,132 meters). At low speeds and altitudes the aircraft had the capability to transport loads weighing a full ton slung under the airframe with cables. The *Drache* operated in a coastal reconnaissance role and carried limited amounts of cargo during World War II. Several Fa-223s survived the war, and one became the first helicopter to cross the English Channel.

Another helicopter, a small, single pilot craft manufactured by Anton Flettner, the Fl-282 *Kolibri* (Hummingbird), also participated in limited combat operations with the German military during World War II. Flettner's rotor system was also a side-by-side design, but his 1939 prototype differed significantly from the Fa-223. Flettner's design, known as a synchropter, placed two shafts in close proximity but angled outward at a significant angle. A synchronizing gear system ensured the exact phasing, or intermeshing, of the overlapping, counter-rotating rotor blades.

In October 1942 two Fl-282 helicopters, accompanied by Kaptain Claus von Vinterfeldt, Flettner's pilot (a man named Fuisting), and another unknown pilot, along with three technicians, arrived in Trieste. The unarmed Fl-282 achieved a maximum speed of 93 mph (150 kph) and a range of 186 miles (300 kilometers). From November 1942 until January/February 1943, one Fl-282 flew reconnaissance missions over the Aegean Sea from the improvised helicopter carrier, Kriegsmarine's minelayer *Drach 50*. Records indicated that the Fl-282 was the first operational shipboard helicopter. The second Fl-232 remained ashore in reserve.

In 1932, under license from Juan de la Cierva, the British Weir Company formed an aircraft department in Scotland. Branching out from autogyros, the company introduced its first true helicopter, the

W-5. Initially a coaxial design, the company revamped the W-5 because of concerns about both stability and control—not to mention the success of the Fa-61. Weir Company engineers reconfigured their prototype as a lateral side-by-side helicopter; it flew successfully in June 1938. The Weir flight control system included cyclic pitch for directional control but provided no collective pitch. The pilot controlled vertical flight by altering the rotor speed, or rpm. The W-5 reached airspeeds of 70 mph in forward flight. The Weir W-6, which first flew in 1939, was a much larger version of the W-5, but further work on the Weir designs was suspended at the outbreak of World War II.

In 1944, Soviet designer Nicolai Kamov launched his first true helicopter, the Ka-8 *Vertolet*. The Ka-8 was a single-seat helicopter powered by a 27-horsepower motorcycle engine that Kamov boosted to 45 horsepower by utilizing alcohol for fuel. On the Ka-8, Kamov introduced the counter-rotating coaxial rotor design that the Kamov Bureau still uses in helicopter production today. Like most helicopter engineers of this era, Kamov used wood laminates for his rotor blades. Only three Ka-8s were built, and records lack any evidence that they saw action during World War II.

Although the German Fa-61 was the first practical helicopter, and the Fa-223 the largest helicopter introduced into limited production during World War II, the most important helicopter advancements took place not in Europe but in the United States. The German concept of counter-rotating rotors mounted side by side did not prove very effective, but the design did, nonetheless, solve the instability of previous helicopters and prompted other inventors to pursue vertical lift technology.

SUCCESSES IN THE UNITED STATES

Concerns over the United States entering the conflict in Europe led military planners to procure new equipment for the armed forces. On April 15, 1940, the Army Air Corps opened bids for its second vertical-flight aircraft. Five companies responded with proposals: three with autogyros and two with helicopter designs. The Platt-LePage Aircraft Company won the bid, mainly because the company proposed a helicopter similar to the Fa-61, already proven in Germany. In 1938, Havilland H. Platt and W. Laurence LePage had formed an aircraft company expressly to build a helicopter. LePage went to Germany to study the Focke-Wulf Fa-61. After LePage witnessed Hanna Reitsch's 1938 performances, he attempted to obtain rights to produce the Fa-61 in the United States, but he could reach no agreement with the German manufacturer. Lacking a contract, the Platt-LePage Company based their helicopter on the side-by-side rotor principle.

The Army designated its project the XR-1 (experimental rotary-wing aircraft model number one), and expected delivery of a prototype in January 1941. The Platt-LePage Company anticipated no problems with the delivery date, as they already had an aircraft, the PL-3, under construction on a speculative basis. The XR-1 (PL-3) prototype weighed 4,730 pounds, with a hightailed fuselage that resembled a flying boat, but with no engine or propeller at the nose. Both the center-mounted Pratt & Whitney R-985 450-horsepower engine and tandem cockpits were fully enclosed. The designers equipped the XR-1 with sliding canopies resembling those of fighter planes of the era. The pilot flew from the front cockpit, and the observer sat in the rear. Plexiglas covered the bottom of the cockpit, allowing forward and downward visibility during landing and also adding to the helicopter's versatility as an observation platform. At the ends of two winglike pylons that extended from each side of the fuselage turned a counter-rotating rotor 30.5 feet in diameter. The cantilevered wings produced some lift in flight, reducing the aerodynamic loads on the rotors. The conventional landing gear consisted of two main wheels and a tail wheel, all free to caster for landing and ground handling.

On May 12, 1941, the XR-1 lifted off the ground for about thirty seconds and soared up about 3 feet. On June 23 the ship made its first untethered flight, but severe vibrations and control problems continually impeded the test program. A young engineer, Frank Piasecki, solved several of the control inconsistencies before he left the company, but other shortcomings plagued the XR-1. On July 4, 1943, several rotor components separated and the helicopter crashed, injuring the test pilot and virtually destroying the craft. Although the aircraft reached 100 mph during one test flight, it remained in the class with Pescara and d'Ascanio.

A few weeks before Pearl Harbor a mock-up of the second prototype, the XR-1A, appeared at the Eddystone, Pennsylvania, test facility. Engineers replaced the old cockpit with a bubble canopy, dramatically increasing visibility. They also swapped the pilot's and observer's stations, a configuration followed in most modern attack helicopters. The engineers also replaced the hubs with a rigid rotor system and installed both new flight controls and a more powerful engine. The new aircraft first flew on October 27, 1943, and in December the test pilot took off from Eddystone, climbed to 300 feet, crossed the Delaware River into New Jersey, and returned to land at the starting point. The high point of the Platt-LePage program occurred on June 20, 1944, when the XR-1A flew cross-country from Eddystone to Wright Field, Ohio. Misfortune continued to bedevil the XR-1, however, and a pinion gear failure on October 26 caused a crash landing that permanently grounded the aircraft. In March 1945 the government canceled its contract with the financially bereft Platt-LePage Aircraft Company. Major advancements in rotary-wing technology came from other inventors who became legends in the helicopter industry.

Born in Kiev, in czarist Russia, on May 25, 1889, Igor Ivanovich Sikorsky, captivated by the drawings of Leonardo da Vinci and the stories of Jules Verne, built a rubber-band–powered model helicopter when he was only twelve. In 1903, Sikorsky entered the Russian Naval Academy in Petrograd, and in 1906 he journeyed to Paris to study engineering. He returned to the Polytechnic Institute of Kiev the next year to continue his studies.

Although the Wright brothers' airplanes captured most of the world's attention, Sikorsky remained focused on a flying machine that could rise vertically from the ground. During a family trip in Germany, he took lift measurements on a 4-foot-diameter helicopter rotor, and with financial backing from his sister, he returned to Paris to study aerodynamics and buy components for his helicopter.

French aviation pioneers attempted to dissuade the young engineer from pursuing vertical flight, but Sikorsky returned to Kiev in 1909 with a three-cylinder 25-horsepower Anzani motorcycle engine and a transmission built to his own specifications. In May, "not yet twenty years old, with a few ideas, no experience, some caution, and, of course, plenty of enthusiasm," he began to fashion a helicopter. With no mechanical assistance he constructed a wooden frame that consisted of a gondola that housed the motorcycle engine on one side and an operator's platform on the other. The transmission filled the center section of the frame and turned a coaxial twin-bladed rotor system. With the coaxial rotors turning in opposite directions, Sikorsky hoped to negate the torque effects created by a single spinning rotor. The upper rotor diameter measured "about fifteen feet" and the lower "about sixteen and a half feet." Sikorsky utilized piano wires and turnbuckles to rig pitch change controls for the rotor blades (Sikorsky 1967, 25–27).

Sikorsky completed construction in July and began testing his new flying machine. Unfortunately he ran headlong into numerous problems that confronted all early helicopter designers. Some problems he solved through experimentation and practical experience; others remained an enigma. The belt that connected the engine and transmission kept slipping, but that was easily corrected. When he applied full throttle the craft vibrated so violently that he was forced to shut down the engine. To solve the vibration problem he removed the rotor blades and balanced them, as well as all the connecting shafts. Sikorsky also discovered that his pitch control system was deficient and that the "contraption" could generate some 357 pounds of lift, about one hundred pounds less than the machine's empty weight. Although he gained a substantial amount of data and pragmatic engineering experience, the youthful inventor realized that his first full-size helicopter would never fly (Sikorsky 1967, 29–30).

In October 1909 Sikorsky dismantled his helicopter and, after viewing Wright biplanes flying in France, completed plans for a conventional airplane while contemplating a new helicopter. In early 1910 he completed construction of "a graceful, but strange looking machine," that "resembled a huge butterfly." The new helicopter proved a valuable testbed, but, again, failed to lift its own weight. Sikorsky realized that he would need a more powerful engine and lighter construction materials before he could build a functional helicopter. He determined that his immediate future lay in the burgeoning aircraft industry. In 1912 the ambitious aeronautical inventor became chief engineer for the aircraft factory of the Russian Baltic Railroad Car Factory in Petrograd, where he designed several successful small airplanes. On May 13, 1913, he became the world's first four-engine pilot when he flew one of his new large, multiengine aircraft. In December 1913 his S-22 began carrying passengers. A bomber version flew in 1914 and went to war with the Imperial Russian Air Force in 1915 (Sikorsky 1967, 30–31).

In 1918 the Bolshevik Revolution forced Sikorsky from his position and drove him from his homeland. A brief wartime engineering effort for the French ended with the November 1918 Armistice. He then booked passage to the United States and arrived in New York on March 30, 1919. After a temporary engineering job with the U.S. Army Air Service, the great Russian aviation innovator taught mathematics to fellow emigres on New York's Lower East Side.

The paucity of funding in the aircraft industry immediately following World War I prevented Sikorsky's return to aviation until March 1923, when he raised financial backing for an all-metal, twin-engined passenger plane. The Sikorsky Aero Engineering corporation began work on a farm near Roosevelt Field, Long Island. To allay financial burdens, the firm's employees collected army surplus materials and parts from junkyards and began producing a series of successful airplanes.

In 1925 the company became the Sikorsky Manufacturing Corporation and introduced several original designs, including the S-34, which provided experience for later amphibians and flying boats. For several years the company produced a variety of flying boats for companies such as Pan American Airways, which specialized in passenger service in the Atlantic and Pacific oceans as well as the Caribbean rim. Sikorsky ultimately drew orders from ten airlines and the U.S. Navy. Charles Lindbergh inaugurated airmail service between the United States and the Panama Canal Zone with Sikorsky amphibians. Financial success allowed Sikorsky to purchase additional land in Stratford, Connecticut, and in 1929 his company became the Vought-Sikorsky subsidiary, later a division of United Aircraft Corporation, which became United Technologies Corporation in 1975.

After the success of his flying boats and amphibians, Sikorsky brought out his drawings of helicopters and renewed his interest in vertical flight. In 1931 he patented the now familiar design of a single large main rotor and small antitorque tailrotor. In 1938 the directors of United Aircraft Corporation rewarded Sikorsky's years of study and research into rotary-wing flight when they agreed to let him, as engineering manager of Vought-Sikorsky, develop a practical helicopter. He immediately began work on a project he called VS-300. The mechanical complexity of his design caused his employees to dub the contraption "Igor's Nightmare," but Sikorsky possessed unbounded confidence in his vertical-lift design and continued the project.

Between the years 1900 and 1939 many inventors experimented with vertical flight. Some of their machines flew, others did not; some of the flimsy machines made only uncontrolled hops into the air, while others sat vibrating themselves to pieces. Some inventors ably demonstrated their genius, but the majority who experimented with rotorcraft did not advance vertical flight. As with fixed-wing aircraft, or conventional airplanes, only a few men discovered, or introduced, innovations on any particular helicopter. When their helicopters failed to meet expectations, many turned to autogyros for vertical flight, or just abandoned their projects altogether. The 1930s proved to be the great decade of helicopter advancement, and although the pioneers of vertical flight made only struggling steps, they laid the foundation for the great technological leaps to come in the succeeding decades.

REFERENCES

Ahnstrom, D. N. *The Complete Book of Helicopters.* New York: World Publishing Company, 1971.

Apostolo, Giorgio. *The Illustrated Encyclopedia of Helicopters.* New York: Bonanza Books, 1984.

Bilstein, Roger E. *Flight in America.* Rev. ed. Baltimore, MD: Johns Hopkins University Press, 1994.

Brian, Marshall. "How Helicopters Work," http://www.howstuffworks.com/helicopter.htm. Accessed September 2002.

Brooks, Peter W. *Cierva Autogiros: The Development of Rotary-Wing Flight.* Washington, DC: Smithsonian Institution Press, 1988.

Brown, David A. *The Bell Helicopter Textron Story: Changing the Way the World Flies.* Arlington, TX: Aerofax, 1995.

Carey, Keith. *The Helicopter.* Blue Ridge Summit, PA: Tab Books, 1986.

Cowin, Hugh W. *Military Helicopters.* New York: Gallery Books; imprint of W. H. Smith Publishers, 1984.

Delear, Frank J. *Igor Sikorsky: His Three Careers in Aviation.* New York: Dodd, Mead, & Company, 1969.

Dowling, John. *RAF Helicopters: The First 20 Years.* London: Her Majesty's Stationery Office, 1992.

Fredriksen, John C. *Warbirds: An Illustrated Guide to U.S. Military Aircraft, 1915–2000.* Santa Barbara, CA: ABC-CLIO, 1999.

Gablehouse, Charles. *Helicopters and Autogiros: A History of Rotating-wing and V/STOL Aviation.* Philadelphia: J. B. Lippincott Company, 1969.

Glines, C. V. "The Skyhook." *Air Force Magazine* 71 (July 1988).

Gregory, H. F. *Anything a Horse Can Do: The Story of the Helicopter.* Cornwall, NY: The Cornwall Press, 1944.

Halcomb, Mal. "The Development History of the Helicopter in Germany from WWI to the End of WWII." *Airpower,* Vol. 37 (March 1990).

Hasskarl, Robert A., Jr. "Early Military Use of Rotary-Wing Aircraft." *Air Power Historian* 12, no. 3 (July 1965): 75–77.

Heatley, Michael. *The Illustrated History of Helicopters.* New York: Bison Books, 1985.

Helicopter's History Site, http://www.helis.com. Accessed June 2002.

Higham, Robin, John T. Greenwood, and Von Hardesty. *Russian Aviation and Air Power in the Twentieth Century.* London: Frank Cass, 1998.

Hunt, William E. *Helicopter: Pioneering with Igor Sikorsky.* Shrewsbury, England: Airlife Publishing, 1998.

Igor Sikorsky Historical Archives, http://www.iconn.net/igor/indexlnk.html. Accessed October 2002.

Keogan, Joseph. *The Igor I. Sikorsky Aircraft Legacy: The Chronology of Fixed-Winged and Rotary-Wing Aircraft of Igor I. Sikorsky and the Sikorsky Aircraft Company.* Stratford, CT: Igor I. Sikorsky Historical Archives, 2003.

Leishman, J. Gordon. "Evolution of Helicopter Flight," http://www.flight100.org/history/helicopter.html. Accessed September 2002.

Liberatore, E. K. *Helicopters before Helicopters.* Malabar, FL: Krieger Publishing Company, 1998.

Pember, Harry. *Seventy-five Years of Aviation Firsts.* Stratford, CT: Sikorsky Historical Archives, 1998.

"Pitcairn PCA-1A." National Air and Space Museum, http://www.nasm.edu/nasm/aircraft/pitcairn_pca.htm. Accessed August 2002.

Ross, Frank, Jr. *Flying Windmills.* New York: Lothrop, Lee, & Shepard Co., 1953.

Saunders, George H. *Dynamics of Helicopter Flight.* New York: John Wiley and Sons, 1975.

Sikorsky, Igor I. *The Story of the Winged-S.* Rev. ed. New York: Dodd, Mead & Company, 1967.

Smith, J. R., and Antony Kay. *German Aircraft of the Second World War.* London: Nautical and Aviation Publishing Company, 1972.

Smith, J. Richard. *Focke-Wulf: An Aircraft Album No. 7.* New York: Arco Publishing Co., 1973.

Spenser, Jay P. *Vertical Challenge: The Hiller Aircraft Story.* Seattle: University of Washington Press, 1992.

Townson, George. *Autogiro: The Story of the Windmill Plane.* Fallbrook, CA: Aero Publishers, 1985.

CHAPTER 2

World War II, 1939–1945

IGOR SIKORSKY COMPLETED his helicopter blueprint and constructed the first VS-300 during the early months of 1939, and on September 14, he sat at the controls when the aircraft made its first vertical takeoff. To ensure stabilization and safety the craft was both weighted and tethered to the ground during initial testing. The first model consisted of a small open cockpit at the forward end of a welded-steel tubing airframe, powered by a four-cylinder, 75-horsepower Lycoming engine that, through a belt transmission, drove both a three-bladed main rotor and an antitorque tailrotor positioned at the end of a narrow tailboom. The pilot sat forward of the engine with only a tachometer, oil pressure gauge, and engine temperature gauge installed in the tiny instrument panel. Sikorsky's control system included a cyclic pitch control for the main rotor and pedals for the tailrotor. For additional yaw stabilization the tailboom also supported a large vertical under-fin.

After its first limited test flights the VS-300 underwent several modifications and captured the interest of the U.S. Air Corps. On May 13, 1940, the revamped VS-300 took flight, unweighted and untethered. An open-framework steel-tube fuselage replaced the enclosed tailboom, and Sikorsky added two outriggers at the aft end of the aircraft. Each outrigger included a horizontally thrusting rotor to augment roll and pitch control. In addition to antitorque control, the inventor expected the three tailrotors to function much as rudder and elevators did on conventional aircraft. Adding more power, a

90-horsepower Franklin engine replaced the Lycoming powerplant. When Captain H. F. Gregory, evaluating the VS-300 for the U.S. Army Air Corps and a novice to helicopter flight, took the helicopter up for about an eight-minute flight, he managed to land near his takeoff point. When Sikorsky and his employees effusively congratulated Gregory on his performance, his companion whispered, "Don't let it go to your head, Frank. They're not congratulating you on your flight, they're just damned glad that you got the thing down again in one piece" (Gregory 1944, 108–109).

Gregory's short flight nevertheless led to an agreement by which the government and Vought-Sikorsky would jointly fund and develop the Army XR-4, in addition to the problematical XR-1. The XR-2 and XR-3 were autogyro designs that soon passed into oblivion. On December 17, 1940, the Department of Defense, Department of Agriculture, Post Office Department, U.S. Coast Guard, and Department of Commerce agreed to provide funding to Vought-Sikorsky for a prototype of the XR-4. The original contract called for a helicopter virtually identical to the VS-300 currently flying, but subsequent test flights required numerous modifications to both the aircraft and contract before final delivery of the XR-4, which only faintly resembled the aircraft specified in the first contract.

Despite being flown successfully by a rank amateur, the machine's flight characteristics still dissatisfied Sikorsky, and he made further incremental improvements. Through a progression of rigorous flights Sikorsky extended the VS-300's time airborne from very short durations to record flight time. On April 15, 1941, Sikorsky set a national endurance record of one hour, five minutes, and fourteen seconds. Until mid-1940 the aircraft had remained airborne for only about fifteen minutes per flight. On May 6, 1941, however, the VS-300 surpassed the world endurance record held by the Fa-61 by remaining aloft for one hour, thirty-two minutes, and twenty-six and one-tenth seconds.

Between 1940 and 1941 the VS-300 crashed several times, and Sikorsky introduced various modifications to improve his helicopter's reliability. In June 1941 he made one of the most important changes by replacing the three rear outriggers with a short vertical pylon that held a single vertical tailrotor. During testing the helicopter had exhibited an unexpected tendency to abruptly pitch nose up at slow hovers. Sikorsky discovered that downwash from the main rotor destroyed the effectiveness of the auxiliary horizontal tailrotors, causing the unexplained maneuver. Not only did the aircraft's controllability improve, but in addition the removal of the added weight improved the helicopter's performance. In December he replaced the previously unsatisfactory main rotor cyclic pitch control with a new system. In another significant alteration he changed the landing gear by replacing the skids with nose and tail wheels. On April 17, 1941, Sikorsky made a successful takeoff from water after fitting pneumatic pontoons under the main wheels. In the VS-300's final form, a 150-horsepower Franklin engine sat in a fabric-covered fuselage with an open cockpit and tricycle landing gear. Sikorsky's determination and flexibility ultimately produced a stable, practical, immensely versatile flying machine that set the standard for subsequent helicopters.

Changes to the XR-4, called S-48 by its manufacturer, under construction mirrored those of the VS-300. Shortly after the Japanese attack at Pearl Harbor in December 1941, Sikorsky employees wheeled the first XR-4 from its secret hangar in Bridgeport, Connecticut. Fabric covered the forward fuselage of the helicopter, the rest of the skeletal tubular framework remaining exposed. A long shaft and gearboxes connected the single tailrotor with the 165-horsepower Warner engine, mounted just behind the two-place cockpit. A cooling cowl directed airflow around the cylinders of the otherwise exposed engine. The engine turned three 18-foot fabric-covered main rotor blades. Just behind the cabin were the engine oil and gas tanks. The pilot's controls consisted of a cyclic stick, directional pedals, and a collective (lift) lever, at the end of which was a twist grip throttle control. Plexiglas windscreens offered excellent visibility as well as some protection for the pilot and observer. Three balloon tires mounted at the ends of hydraulic shock struts comprised the main tricycle landing gear. A tiny wheel attached to the tail supported the rear of the aircraft. On January 14, 1942, before a crowd of government officials and company dignitaries, the XR-4 lifted into the air for the first time.

Testing on the final version of the VS-300 continued for several months before Vought-Sikorsky officially retired the helicopter. The VS-300 flew throughout 1942 despite the fact that the development of Sikorsky's first production helicopter, the S-47 or XR-4, was well underway. In 1943, Vought-Sikorsky delivered the VS-300 to the Henry Ford Museum in Dearborn, Michigan, where it is housed today.

The S-47/XR-4 underwent several minor modifications and test flights before the first prototype was accepted by the U.S. Army Air Corps on May 30, 1942. Sikorsky expanded the instrument panel to include a rotor tachometer, a blade pitch indicator, and the basic flight instruments found in most airplanes of the era. He also added Plexiglas in the forward area of the floor to increase downward visibility. During an April test flight the pilot hovered up to a pole on which the ground crew had attached a brass ring. As if he were on a carnival carousel, the pilot slipped the pitot tube through the ring and returned to land the prize in front of the S-47's inventor. On another flight, to prove the aircraft's versatility, the XR-4 lifted a net bag full of eggs slung under the aircraft by a thin rope. The pilot circled around and set the net back on the ground so gently that none of the eggs were broken. An engineer took one of the eggs and smashed it on a rock to prove that they were not hard-boiled. Demonstrating the capability to communicate with a hovering helicopter, the observer lowered a telephone handset to an officer on the ground and carried on a short conversation; radios, of course, soon replaced the telephone. Sikorsky pilots demonstrated the singular rescue capability of the helicopter by letting down a 25-foot rope ladder to the ground, picking up a man on the ladder, and flying away with him. Reversing this operation, a helicopter could deliver critical personnel to isolated units. In addition to military applications, the helicopter could be used to scout for forest fires and lower firefighters into otherwise inaccessible locations, possibly quenching a fire before it became an uncontrollable conflagration.

With a gross weight of 2,700 pounds, the XR-4 cruised at 70 miles per hour at 5,000 feet and, with certain modifications, demonstrated its usefulness to the U.S. Navy. Fitted with inflatable rubber floats, which were described as "large hotdogs," the helicopter made several takeoffs and landings from both land and water, including the choppy surface of Lake Erie. Navy officers immediately devised several uses for the helicopter: search and rescue, convoy and antisubmarine patrol (without an observer the machine could carry a depth charge), and adjusting naval gunfire. On May 6 and 7, 1943, in Long Island Sound, an Army test pilot made several takeoffs and landings aboard the USS *Bunker Hill*, a freighter modified with a landing platform between the superstructure and cranes. The helicopter's 38-foot-diameter rotor left little room for error, and approaches had to be made from the side of the ship, but the pilot successfully negotiated twenty-four landings and takeoffs from the ship, both anchored and underway. After the demonstrations, the pilot ferried several dignitaries from the ship to the Stratford airport—another first for the Sikorsky helicopter, ship-to-shore transfer of passengers by aircraft. In July the XR-4 and an R-4 conducted a successful three-day trial of shipboard landings utilizing a platform built on the stern of the USS *James Parker,* a luxury liner converted to a fast troop transport. These fast ships steamed alone, not in convoy, and military planners hoped that the R-4, armed with rockets, could protect the ships from submarines. Unlike the previous shipboard landings, these were conducted on the open ocean to a rolling ship and with 40-knot winds. With its ability to hover and land aboard ships, the XR-4 seemed perfect for the Navy as well as the Army. The Navy designated its first version the HNS-1.

On the 761-mile cross-country trip from the Vought-Sikorsky Stratford, Connecticut, plant to Wright Field, Ohio, and successive flights, the XR-4 broke all existing records for endurance, altitude, and airspeed. The XR-4 flew 100 hours without major incident, climbed to a demonstrated service ceiling of 12,000 feet, and approached a maximum airspeed of 90 miles per hour. U.S. Army test pilots took the helicopter to the very limits of its flight envelope, effectively demonstrating both the military and civilian versatility of the helicopter. On January, 5, 1943, the Army ordered the production of twenty-nine XR-4s.

On May 16, 1943, one of the first production R-4s became the first helicopter to carry U.S. mail. Initiating a twenty-five-year celebration of the mail service, an R-4 lifted off from the Capitol terrace and delivered a small package of air mail to a waiting airliner at the Washington, D.C., Municipal Airport. The helicopter accomplished the historic flight through miserable weather conditions in which no conventional airplane could fly—fog and drizzle that reduced visibility to an eighth of a mile. A few weeks later Lowell Thomas, a nationally known news commentator, made the first radio broadcast from a hovering R-4.

Although it appeared much the same as the prototype, the production R-4 included several improvements. The exhaust pipe vented out the side instead of the bottom of the aircraft, which eliminated accidental fires when hovering over combustibles such as dry grass and leaves. The tail wheel was moved farther back on the tailboom to protect the tailrotor during an autorotative landing. Engineers repositioned air ducts to draw engine cooling air from the front of the helicopter instead of the rear. A Warner 180-horsepower air-cooled engine replaced the previous 165-horsepower model. A previous 1-foot extension to the rotor blades had expanded the rotor diameter to 38 feet, thus providing more lift. Increasing the fuel capacity by 5 gallons added to the machine's range, and the installation of a radio provided air to ground communications lacking previously. The aircraft bore the traditional U.S. Army olive drab paint on the upper surfaces; battleship gray covered the underside.

Before deploying the R-4 worldwide, military planners directed that the helicopter be tested in extreme weather conditions. On November 6, 1943, an R-4 began a long trip to Ladd Field near Anchorage, Alaska. At Bridgeport, Connecticut, mechanics disassembled an R-4 and loaded it aboard a C-46 transport aircraft. After arriving in Alaska the R-4 was reassembled and went through a succession of cold weather tests. Nicknamed *The Arctic Jitterbug,* the helicopter, after being serviced with oil and hydraulic fluid designed for cold weather operations, performed satisfactorily in every test. During the program test personnel fabricated a metal frame that they attached to the side of the R-4. To the frame they attached a standard Army litter, protected by a canvas shroud. Several flights concluded that casualties, placed in sleeping bags in cold weather, could easily be evacuated from the battlefield to field hospitals. Program managers rapidly discovered, however, that frost deteriorated the fabric covering the rotor blades, and they ordered special covers to protect the blades from the elements. Pilots also determined that flights into icing conditions would shred the rotor blades. Department of the Interior personnel observing the helicopter trials quickly recognized the usefulness of such a machine in wildlife and forest management. Although the R-4 had not seen combat, it had, nonetheless, flown under adverse conditions in a combat theater of operations (Gregory 1944, 201–206).

Continued testing of the XR-4 revealed a few problems with some components, but close cooperation between Vought-Sikorsky engineers and Army personnel overcame all obstacles. The helicopter's transmission overheated and required replacement gears more often than expected. Only four days before General Henry H. "Hap" Arnold, chief of Army Air Forces, arrived on a special trip to view the capabilities of this marvelous new machine, the mechanics discovered that the helicopter required a new transmission. No spare was available at Wright Field. After a frantic telephone call to Connecticut, company engineers scrambled to gather the necessary machinists over the Fourth of July holiday. Rushing the large box of hastily manufactured gears to the airport, Sikorsky personnel, as a war-time emergency measure, replaced two irate TWA passengers with the repair parts, and the box rode in the passenger cabin to Dayton. On July 7, 1944, after a hurried overhaul, the XR-4 performed flawlessly and so impressed Arnold that he ordered procurement of the larger and more powerful Vought-Sikorsky XR-5, the first helicopter ordered in quantity by the U.S. armed forces.

As early as 1942, Sikorsky constructed a detailed mock-up of a more advanced rotorcraft, which became the R-5. Although still only a two-seater, the R-5 was much larger, more powerful, and more capable than the R-4, which, in many cases, was relegated to pilot training. Planned with universal capabilities, the mock-up included attaching points for four litters (two on each side) carried in protected capsules and provisions for installing bomb racks on the R-5's belly. The preliminary plans met the Army requirements for an airspeed in excess of 125 mph, efficient operation above 5,000 feet, endurance over three hours, and a useful load of more than 1,100 pounds. On June 13, 1943, the War Department sent Sikorsky a letter stating that the government requested four XR-5s, two U.S. and two British, paid for by Lend-Lease funds.

SIKORSKY XR-5 AND XR-6

The XR-5 differed radically from its predecessor. Wartime shortages, for example, forced Sikorsky engineers to construct some components of molded plastic–impregnated plywood instead of aluminum. The three-bladed rotor system measured 48 feet in diameter, with blades constructed of laminated plywood over a steel tube spar and canvas trailing edges. A nine-cylinder, air-cooled 450-horsepower Pratt & Whitney Wasp Junior radial engine, mounted vertically, powered the new helicopter. Three sections embodied the XR-5's 40-foot fuselage. An aluminum alloy frame, enclosed by Plexiglas and mounted on an aluminum monocoque floor, made up the cockpit area. Tandem seating, with the observer in front and pilot behind, offered almost unlimited visibility. Dual controls allowed either to fly the aircraft. The center section, containing the engine, was constructed of welded steel tubing covered with plastic-impregnated plywood. The aft section, a wooden monocoque tail cone, supported the solid wood, 7-foot tailrotor and its driveshaft. The two-man crew sat forward of the main wheels of a conventional landing gear.

On August 18, 1943, the XR-5 first took flight, but it experienced severe vibrations and unacceptable bending of the main rotor blades. After stiffening the ends of the blades and adding balance weights to them, the test engineers tried again, achieving more success. On September 13 the XR-5 lifted ten people to a hover, two in the crew positions and four hanging onto each side of the landing struts. Vought-Sikorsky produced the R-5 in substantial numbers for both the Army and Navy (HO5S), and several of them saw military service in the Pacific and China-Burma-India (CBI) theaters during World War II. The company also manufactured sixty-five units of the company-funded R-5B, with a cockpit modified to seat three passengers behind a forward pilot position.

Engineers and pilots quickly recognized the inadequacy of the power-to-weight ratio of the R-4, especially after the addition of radio equipment and bomb racks. This realization engendered ideas for a more powerful, streamlined helicopter, which resulted in the XR-6 design. Lack of financing almost killed the XR-6 before the project began, but the U.S. Navy saved the new helicopter. Navy planners envisioned future requirements for a still larger and more powerful helicopter and agreed to fund half the development costs of the XR-6. One wag described the streamlined helicopter as "a gourd with rotors and wheels" (Gregory 1944, 221).

As with its predecessor, three main sections—cabin, center, and tailboom—composed the XR-6. The ship differed, however, in that lightweight alloys and newly discovered composites lightened and strengthened the airframe. An aluminum alloy monocoque floor ran back from the cockpit and under the center section, much like the keel of a boat. An aluminum frame, encapsulated with Plexiglas and fiberglass, made up the forward cockpit area. Inside the molded plastic center section, a steel alloy tubular frame supported an air-cooled, six-cylinder 245-horsepower Franklin engine, along with the transmission, fuel, and oil tanks. Designed for increased streamlining, the engine intakes and exhaust stacks were flush-mounted to the fuselage. Aluminum alloy bulkheads, covered with a tough magnesium skin, made up the tailboom, which also housed the driveshaft for the three-bladed, 7-foot, 10-inch tailrotor. The helicopter rested on a conventional landing gear, with a small nose wheel to protect the craft in the event of a nose-over and to improve forward taxiing. The pilot and observer sat side by side with controls identical to those of the R-4. Directly behind the crew seats a shelf held standard military radio equipment. Considerate designers also included special pocket containers for maps and other aviation paraphernalia. The rotor blades were the same size as on the R-4, and interchangeable, but the R-6 sported a newly designed, streamlined rotorhead.

Improved performance also characterized the R-6. Airspeeds of over 100 mph and endurance of more than five hours gave the R-6 the longest range of any helicopter up to that time. With operating weights around 2,600 pounds the R-6 could cross the Rocky Mountains with room to spare. The craft could carry an enclosed litter on each side, or in lieu of litters, bomb racks.

The XR-6 made its first flight on October 15, 1943. Extreme vibrations, control feedback, and other issues limited the flight time and restricted the altitude to about 1 foot, challenging the designers for solutions. As with the R-4 the transmission overheated, especially in hovering flight. Despite Sikorsky engineers' best efforts the vibrations continued to rattle the helicopter, increasing with airspeed, but those associated with the new rotorcraft considered these minor growing pains. On November 27, in spite of some problems, the XR-6 lifted six personnel, including the crew, and hovered for several minutes. By early 1944 engineers managed to smooth out most of the vibrations, and test pilots conducted flights into snow without experiencing any deterioration of the rotor blades. Encouraged, pilots began to believe they could fly in "helicopter weather," conditions in which most other aircraft were grounded.

By the spring of 1944 the XR-6's performance justified the time and effort expended in its construction. On March 2, Colonel H. F. Gregory, USAAF, carrying a civilian observer, flew the ship nonstop from the Washington National Airport to Wright Field, Ohio. The trip of some 387 miles took four hours and fifty-five minutes, unofficially breaking three world records, including the longest nonstop flight of a helicopter. The flight's ground speed, against head winds, also bettered any previous mark. The XR-6's heated, soundproofed cockpit, which reduced engine and transmission noise significantly, also impressed anyone who flew the helicopter.

Although shortcomings and limited capabilities hampered the R-4, R-5, and R-6's performances, the government purchased 130 R-4s and more than 300 R-5s and R-6s, and each type of helicopter commanded some historic milestones serving with the U.S. Army, Navy, and Coast Guard, and the British Royal Navy and Royal Air Force, which dubbed their R-4s the "Hoverfly." On January 3, 1944, an R-4 flew the first helicopter mercy mission, rushing blood plasma to the survivors of an explosion aboard the USS *Turner*. During the night of December 31, 1943, a mysterious explosion shook the destroyer, killing several of the crew and injuring many others. Coast Guard ships plucked survivors from the icy water, evacuated the injured to emergency hospitals set up along the beach at Sandy Hook, New Jersey, and sent out an emergency call for blood plasma. Commander Frank A. Erickson, commanding officer of the Coast Guard Air Unit at Floyd Bennett Field, Long Island, flew a Coast Guard HNS-

1 (R-4) through a swirling snowstorm to land on the tiny lawn at Battery Park in lower Manhattan, picked up the crucial plasma, and fourteen minutes later delivered it to the medical personnel at Sandy Hook. No other means of transportation could have responded with the life-saving plasma so rapidly.

Coast Guard R-4s also assisted in calamities affecting civilians. On April 1, 1944, Lieutenant (jg) W. C. Bolton, on a routine patrol from Floyd Bennett Field, spotted a lone individual on an isolated sand bar in Jamaica Bay. He swooped down and rescued a fifteen-year-old boy who had been marooned on the bar by incoming tides. Also in April a fire in some ties on a Long Island railroad threatened to destroy the trestle over Jamaica Bay. Firemen fighting the blaze could not reach the fire with the necessary equipment. A Coast Guard helicopter from Floyd Bennett Field flew to the scene, hovered over the blazing trestle, and lowered hand extinguishers to the fire fighters.

In November 1944, while scouting over a lake near Orlando, Florida, for a submerged P-51 fighter, U.S. Army Captain McGuire, flying a Sikorsky XR-5, received a radio communication that a C-47 cargo plane had crashed 50 miles away. McGuire flew to the area to search for the crew of the downed aircraft. He picked up an injured chaplain who had been aboard the C-47 and evacuated the injured man to a base hospital.

The first combat rescues of downed airmen and wounded soldiers took place in the CBI Theater by Sikorsky R-4s. In April 1944, Lieutenant Carter Harman proved that the noisy, ungainly R-4 possessed the capabilities to save lives, when he rescued a trio of injured British soldiers and their U.S. pilot, T/Sgt. Edward Hladovcak, whose liaison plane had been shot down behind enemy lines. Despite the high altitude, humidity, and tropical temperatures that reduced the helicopter's performance, Harman extracted the four, two at a time, out of a rice paddy almost under the noses of the Japanese—a unique rescue, because the R-4 was designed to carry only one passenger.

During May 1944, R-4s participated in several other evacuations in the CBI Theater. During an Allied airborne invasion of Burma paratroopers suffered a number of casualties, some combat casualties, some not. U.S. Army pilots evacuated a number of the casualties, including litter patients strapped to the outside of the helicopters.

At midnight on January 23, 1945, the Tenth Air Force Air Jungle Rescue Detachment received a call that an enlisted man at a weather station, high in the Naga Hills in northern Burma, had accidentally shot himself through the hand. The wound had become infected and required prompt medical attention. The station commander wondered if the eggbeater could land at high altitude and lift the man out. Captain Frank W. Peterson flew an R-4 to the mountain station and, by making a running takeoff, managed to fly the injured man to a hospital.

Lieutenant Raymond F. Murdock, piloting an R-4, conducted one of the most dramatic rescues in the CBI Theater when he plucked a severely injured pilot from the dense Burmese jungle. On March 19, 1945, an Air Transport Command C-46 had iced up and crashed while attempting to carry supplies over the Himalayan Mountains. Captain James L. Green and a Naga tribal chieftain who claimed to know the location of the downed crew took off in a Fairchild PT-19 trainer from Shingbwiyang Airfield, Burma. After two hours of fruitless searching Green turned back to the airfield. Only five miles short of the field the PT-19's engine quit, and Green crashed in the 150-foot-tall trees covering the last steep ridgeline before the airfield. At dusk another aircraft spotted the tangled wreckage of Green's plane. The next day a ground rescuer party began a two-day trek to slash their way to the crash site. The rescue party located the seriously injured Green and the dead tribal chief. The flight surgeon who accompanied the rescue attempt determined that Green could not survive the trip back to Shingbwiyang. The doctor believed, however, that with air-dropped medical supplies he might keep Green alive for a few days. A message went out to the Air Jungle Rescue Detachment at Myitkyina, requesting a helicopter rescue; no other aircraft had any hope of landing near Green. Volunteers hauled hand tools, power saws, and dynamite up the steep slope and, in two weeks, hacked out a hole in the trees and built a small platform on a ledge jutting out from the ridge.

A severe rain squall on the night of April 3 threatened to terminate the entire operation, but at daylight on April 4 the bamboo-reinforced landing pad remained intact. Dense fog kept Murdock grounded until 10:00 A.M., when he cranked the R-4 and took off for the rescue, hoping that the increasing temperatures would not degrade the helicopter's performance to the point that it could not lift two people out of the small landing zone. To complicate the situation further, gusty winds buffeted the helicopter as he approached the ridgeline. Murdock thought, "This is going to be like landing in a windy well," but he managed to hover the R-4 down through the trees and onto the pad. When he reduced power the helicopter began to roll off the pad. Several men rushed out and held the R-4 in place while others carried Green to the waiting helicopter. They strapped Green on a reclining board that replaced the right seat, and Murdock prepared for the most harrowing phase of the mission. He cranked the engine and twisted the throttle to maximum power. When he lifted the collective lever the R-4 jumped about 4 feet off the ground but would climb no higher with both men aboard. Murdock swallowed hard and turned the helicopter downhill and started forward. With the combination of max power, forward airspeed, and down-sloping terrain Murdock barely cleared the trees and deposited Green at the Shingbwiyang hospital. The little helicopter failed, however, to fly back to its base at Myitkyina. The engine, overdue for replacement, quit en route, forcing Murdock to make an emergency landing on the Burma Road. The R-4 finished the trip on the bed of a truck. Sikorsky helicopters continued their valuable service rescuing soldiers and airmen in the CBI Theater throughout the remainder of World War II (Glines 1988, 109–110).

In November and December 1944, RAF Wing Commander Brie revealed that Sikorsky helicopters also had served with the armed forces of Great Britain in the Atlantic Ocean and Europe. The British lashed R-4Bs to the decks of warships convoying cargo vessels to England. The British utilized the helicopters to search for U-boats and for survivors in the event of submarine attacks. He also related that R-4Bs equipped with pontoons had rescued several downed pilots from the English Channel and from mudflats along the English Coast.

On January 30, 1945, a Coast Guard R-4 flew a unique mercy mission by landing an inhalator aboard a merchant ship in New York harbor. A crewman had fallen overboard into the icy waters and first-aid efforts to revive the seaman had failed. Harbor ice prevented the ship from transporting the man ashore, and the captain radioed for assistance. Heeding the call, in less than a half-hour a Coast Guard helicopter hovered over the ship and lowered the inhalator to the deck of the ship.

In April 1945 a U.S. Air Transport Command C-54 cargo plane ferried a Coast Guard pontoon-equipped R-4 more than 1,000 miles from Floyd Bennett Field to Labrador for another rescue. Nine crewmembers of a Royal Canadian Air Force plane that had crashed in the inaccessible Labrador wilderness remained stranded and required prompt succor. On May 2, after the ground crew reassembled the R-4 at Goose Bay, the pilot battled winds, snow squalls, and blinding sun to reach the crash site and, on successive flights, managed to rescue all the downed airmen. The pilot reported a severe sunburn after hours of flying in the brilliant glare of the reflected sunlight.

Sikorsky helicopters also supported the logistical component of war toward the end of World War II. On May 22, 1945, the U.S. Navy revealed that helicopters had been assigned to floating depots in the Pacific Theater. Floating air depots on modified Liberty ships, equipped with aircraft repair and maintenance shops, accompanied Allied thrusts into Japanese-held territory. Sikorsky R-4B helicopters operated from tiny flight decks aboard these ships, ferrying parts ashore or to other ships to keep combat aircraft flying.

Toward the end of World War II, helicopters aptly demonstrated the impact that rotorcraft would have on the civilian world after hostilities ceased. In July 1945 a Sikorsky helicopter sprayed insecticide at the Yale Bowl. Prior to the first outdoor concert of the New Haven Symphony Orchestra at the Yale Bowl, a Coast Guard helicopter sprayed the bowl and adjoining acres with DDT, the war-developed insecticide. The Connecticut Agricultural Experiment Station and the USDA Bureau of Entomology and Plant Quarantine cooperated with the Coast Guard to complete the project. During the same month, U.S. and Mexican Army pilots, flying Sikorsky helicopters, flew several scientists over the Paricutin volcano near Mexico City, which had become active in 1943. The missions offered opportunities both to observe a developing volcano scientifically and to test flying conditions. Although the flying closely approximated that experienced in the CBI Theater, U.S. Army Captain Colchagf related that the pilots discovered information of vital interest to the Engineering Department at Wright Field. Professor L. C. Graton, of the Harvard Geology Department, concluded that there was no doubt that observations from the slow-flying helicopters afforded the scientists knowledge of the volcano's activities that could never have been obtained from the ground or from conventional aircraft.

On November 29, 1945, an S-51 (R-5) performed another aviation first. Responding to a distress call to assist the Coast Guard in rescuing the crew of an oil barge breaking up in heavy weather, Sikorsky test pilot Dimitry "Jimmy" Viner and an Army Air Force officer took off from the Sikorsky plant despite strong winds and heavy rain. They flew low over the choppy waters off Fairfield, Connecticut, and spotted the sinking barge. Viner hovered over the stricken craft while the Army officer employed the hoist to lift the two crewmen to safety, thus performing the first hoist rescue in history.

FRANK NICHOLAS PIASECKI

Although credited with creating the first successful helicopter of the United States, Sikorsky did not work in a vacuum. Several other young and ingenious helicopter designers labored throughout the 1940s to bring their perceptions of vertical flight to fruition. Frank Nicholas Piasecki, Arthur Middleton Young, Stanley Hiller, Jr., and Charles Kaman vied for recognition and eventually unveiled successful helicopters.

On April 11, 1943, Philadelphia native Piasecki flew the second successful prototype helicopter in the United States. Born in Philadelphia in 1919 the son of an immigrant Polish tailor, Piasecki earned degrees in aeronautical and mechanical engineering from the University of Pennsylvania. After graduation he worked for both the Kellett Autogyro Company and Platt-LePage Aircraft Company as an engineer. With Harold Venzie, he formed the P-V Engineering Forum in 1940. Piasecki set to work and designed his first helicopter, a tiny single-seater called the PV-2, which he believed would revolutionize private transportation. He constructed most of the first prototype in his garage and crashed it on its initial flight. Of note, Piasecki was the first in the United States to receive a pilot's license to fly helicopters before qualifying to fly conventional aircraft.

On account of wartime shortages Piasecki scrounged many of his parts from crashed aircraft and auto salvage yards, but he equipped his first helicopter with a dependable and lightweight four-cylinder, 90-horsepower air-cooled Franklin engine, mounted vertically under the rotor blades. The entire fabric-covered craft, including pilot and fuel, weighed just over 1,000 pounds. Elliot Daland, a PV engineer who had also worked for Platt-LePage, incorporated an innovative feature into the rotor system. Although Daland followed conventional blade construction utilizing a steel tube for the main spar and covering wooden ribs with fabric, he included internal adjustable weights to balance each blade dynamically and adjust its center of gravity, thus reducing unwanted vibrations. The three-bladed rotor system measured 25 feet in diameter, but Daland ingeniously designed two blades to fold back, decreasing the PV-2's length to only 22 feet. With the blades folded for storage the helicopter measured about 9 feet in height and 8 feet wide, allowing the craft to fit into a garage or be towed behind an automobile. The pilot sat in a conventional seat mounted in a rounded, glass-covered nose compartment. Piasecki's team installed conventional helicopter controls: a cyclic pitch control stick, a collective (lift) lever, and pedals that controlled the single vertical tailrotor.

To bring attention to the little PV-2's versatility Piasecki carried out a novel stunt. He took off from the driveway of a private home in Falls Church, Virginia, flew a short distance to a filling station, and landed for gasoline. He spent an "A" ration stamp for three gallons of gasoline. The astonished attendant put in the gasoline, cleaned the helicopter's windshield, and watched as Piasecki took off for a nearby golf course. He landed by the first tee, retrieved his golf clubs from the small baggage compartment, and proceeded to tee off for a round of golf.

Flying the PV-2 rekindled Piasecki's interest in military transport aircraft, and he designed a much larger overlapping tandem rotor helicopter. Chided by a Senate panel chaired by Harry S. Truman for its lack of interest in the obvious military potential of the helicopter, the Navy Department searched for a new machine. In January 1944 the Navy awarded Piasecki a contract, and he began construction of a flying prototype of the PV-3. At Morton, Pennsylvania, on March 7, 1945, the PV-3 (XHRP/HRP-X) *Dog House*, the world's first successful tandem-rotor helicopter, took to the air for the first time. XHRP indicated Experimental Helicopter Transport; HRP-X indicated that this was not a U.S. Navy aircraft but a company prototype. The official Navy prototype, called the *Rescuer,* flown later, fell under the XHRP-1 designation. Piasecki dubbed the ship the *Dog House* because dogs were used in many instances as guinea pigs.

Flown by a crew of two in tandem seating, equipped with eight passenger seats, and powered by a 600-horsepower Pratt & Whitney engine, the PV-3 was the largest helicopter flown to date. Piasecki engineers mounted the single engine in the rear section of a lightweight steel-trussed fuselage that, through driveshafts, powered a gearbox amidships; the gearbox, in turn, drove reduction gearboxes installed under each of the three-bladed rotor systems. Using lessons learned from his days at Platt-LePage, Piasecki mounted the rear rotor hub higher, so that downwash from the front blades would not aerodynamically interfere with those at the rear. Initially a Plexiglas windscreen protected the pilot, who sat in the forward seat, but no skin covered the airframe. Later, the installation of a fabric skin greatly increased the ship's forward airspeed. The first prototype almost suffered a catastrophe because the transmission, manufactured to automotive tolerances, could not satisfy the rigid demands on a helicopter's drivetrain. When the gearboxes overheated on a test flight, Piasecki did not want Navy officials to learn of the problem. He sent his flight engineer out to buy ice and soda pop, which they poured over the transmission to cool it down.

The XHRP-1 offered more versatility and cargo capacity than any previous helicopter. Removing the passenger seats from the *Dog House* made possible space for a ton of cargo or the installation of six standard litters. In June 1946 the U.S. Navy and Coast Guard ordered twenty units for search and rescue and utility transport duties, with the last delivered in 1949. The USCG received only three aircraft, while twelve of the machines eventually went to the U.S. Marines for assault training. When the military retired the HRP-1s from service in the early 1950s, six appeared on the civil aviation registry. The distinctive bent fuselage shape earned the PV-3, and later Piasecki models, the sobriquet "Flying Banana." A company comic joked that the chief engineer called Piasecki on the telephone and told him that the production team mistakenly attached two helicopter front assemblies together, but "the good news is that it flies."

FROM MILITARY TO COMMERCIAL HELICOPTER

Born of wealthy Pennsylvania parents, Arthur Young graduated with a degree in mathematics from Princeton University in 1927. More interested in philosophy than mathematics, however, he searched for a project to occupy his time and discovered a book by Anton Flettner that stimulated his interest in helicopters. In 1928, Young set up a workshop on the family estate and began experimenting with a variety of helicopter models. His first helicopter consisted of rubber bands, carved wooden propellers, and balsa wood strips, all of which he bought at a toy store. The first model, which was 6 feet in diameter, took off and flew about ten seconds before it crashed. During the late 1920s and early 1930s, Young designed and redesigned several models, making use of everything from rubber bands to electric motors to power his experimental craft. Constant failures beset him, and after a large model powered by a 20-horsepower outboard motor crashed, he discarded Flettner's ideas of fixed rotor blades and pursued his own aerodynamic calculations.

Following his own concepts Young invented a semirigid rotor system. The main rotor blades were attached to the hub in such a way that they flapped up and down like a seesaw, or teeter-totter. He attached something new to his teetering rotor system, a stabilizer bar. The bar had weights attached to each end and was directly linked to the rotor blades through the pitch control linkages. Young discovered that when the rotor was disturbed in pitch or roll, the gyroscopic inertia of the bar, by cyclic pitch control inputs, effectively dampened rotor disturbances, thus stabilizing the rotor system. Impractical on his models, Young replaced the bar with a flywheel arrangement that could be remotely controlled during flight.

Resolved to build a full-scale helicopter, Young sought to interest aircraft manufacturers in his concepts. No one seemed interested until, by word of mouth, accounts of his experiments reached aviation entrepreneur Lawrence "Larry" D. Bell. Captivated by airplanes since boyhood, Bell had begun a career in the aircraft industry working for Glen Martin at Martin Aircraft. He then moved on to Consolidated Aircraft as a design engineer and became general manager at the Buffalo, New York, plant. He left Consolidated Aircraft and formed his own company, Bell Aircraft, in the late 1930s. Government "Victory loans" and lucrative contracts during World War II guaranteed Bell Aircraft's success. The company produced the P-39 *Aircobra* and P-63 *King Cobra* for the United States, and the P-400 for the Free French, British, and Soviet air forces. Late in the war Bell Aircraft also built B-29s under contract at a new plant constructed near Marietta, Georgia.

After reports of Young's success at flying a remotely controlled helicopter model in and out of his barn workshop, Bell invited Young to demonstrate his model at the Bell plant. During a trip to Germany, Bell had witnessed several of the Fa-61's flights, and the idea of a helicopter intrigued him. On September 3, 1941, Young maneuvered his remotely controlled helicopter model around the crowded plant and then showed a film about his experiments. After the demonstration Bell and Young made an agreement to build two full-size helicopters, Bell to provide funding and Young the design.

On November 24, 1941, Young and his assistant, Bartram "Bart" Kelley, arrived at the Buffalo plant to supervise the initial construction of two prototypes stipulated in the contract between Bell and the inventor. The Japanese attack on Pearl Harbor and a few misconceptions, however, slowed initial research. Wartime production monopolized Bell's interest, and his concern about what happened when a helicopter engine failed led to a halt in the funding of Young's project. To allay Bell's concerns, Young placed a raw egg in his model and flew it up to the plant ceiling. Young then cut the power to the helicopter and completed a smooth autorotational landing to the floor without breaking the egg. Convinced, Bell agreed to proceed. The two men established such trust between each other that Young signed over his patents to Bell Aircraft. On November 10, 1941, Young filed for a patent on his rotor system with the stabilizer bar; the final patent approval from the U.S. Patent Office was issued on September 11, 1945.

Young decided that the only way to move forward with his helicopter was to move completely out of the tumultuous aircraft production facility at Buffalo. He found a vacant automobile dealership in Gardenville, New Jersey, and convinced Bell to provide a staff of thirty-two engineers, machinists, sheet metal workers, and mechanics to build the Bell Model 30 helicopter, known to history as Bell Ship 1. Young also convinced Bell to allow him freedom from company administration while developing the first prototype. In June 1942, Young and his team moved into the new facility and commenced work on the helicopter. They expanded Young's demonstration model by six times to construct Ship 1, which took about six months to build. A variety of construction materials went into Ship 1. Plywood beams and steel tubing formed most of the skeletal fuselage and riveted magnesium the tail cone. Young constructed the rotor blades of a laminate of fir and balsa woods, reinforced by a steel bar down the leading edge of each blade. A cowling partially protected the open cockpit, with everything else exposed. A 165-horsepower Franklin engine, mounted vertically, drove both the 32-foot-diameter main rotor and the antitorque tailrotor through a geared transmission and driveshafts. The Gardenville plant copied Young's model transmission exactly, because no one at the Bell plant knew how to design and build a transmission from scratch. Young also fitted Ship 1 with a unique control system—it lacked rudder pedals—all control functions being done with the hands.

Like other inventors Young encountered impediments to progress, both mechanical and human. Ship 1 initially perched on a landing gear consisting of four 12-foot spider legs welded to a tubular fuselage. The legs, with shock absorbers about 4 feet above the ground, were designed to prevent the helicopter from inadvertently tipping over during hover tests. Because of the legs' length Ship 1 would not fit through the garage door, and mechanics had to remove and reattach the landing gear each time the helicopter was rolled out for tests.

On December 18, 1942, Young's team rolled the ship out of the garage for its inaugural test flight, but winter weather postponed the helicopter's flight: the engine became too cold to start, but a more powerful battery cranked the engine and on December 29 the helicopter lifted off on its first tethered hovering flight. Bell had no designated test pilot, so Young climbed into the pilot's seat and flew the first test flight himself. Up to that point he had flown nothing but his remote-control models.

In late January 1943 a needless mishap severely damaged Ship 1. Robert Stanley, Bell Aircraft's chief pilot (and later chief engineer), came to Gardenville and demanded to fly the helicopter. No one felt empowered to challenge the company's chief pilot, so he climbed into the pilot's seat and started the engine. Stanley, like many airplane pilots, disdained helicopters and failed to fasten his safety belt. When Stanley lifted Ship 1 off the ground, he worked the collective control rapidly like the pump handle that it resembled. The helicopter reacted violently, and he was tossed up, out of the seat, and into the whirling rotor. Fortunately for Stanley, the steel stabilizer bar missed him, but the wooden blades whacked him, breaking his arm. The impact tossed him several feet, and he landed in a heap. In the ensuing crash of the pilotless helicopter both the rotor blades and the tailboom were so badly damaged that they had to be replaced, delaying the test program for several weeks. Chagrined at his lack of forethought, Stanley called Bell to inform him that he had delayed the helicopter project.

Bell sent Floyd Carlson to New Jersey as the designated Model 30 test pilot and to help coordinate the project. The Bell team reconstructed Ship 1 and began a series of tethered flights to work out some of the ship's idiosyncrasies. On June 26, 1943, the helicopter finally made its first untethered flight.

As flight tests progressed Young's engineers and pilots discovered a number of unforeseen problems. Airspeeds above 30 mph induced extreme vibrations that shook the helicopter, making it almost uncontrollable. The engineers discovered that the vibrations emanated from flimsy rotor blades, and a stiffening yoke eliminated the vibrations. Undue wear on transmission gears also delayed flight tests. Every hour or two of flight time necessitated a complete tearing down of the transmission and rotors to determine where the wear had occurred. As the engineering team overcame one problem after another, Ship 1 exceeded 70 mph in forward flight.

While the test pilots flight-tested Ship 1, the mechanics and machinists constructed Ship 2, which differed radically from her predecessor. According to Bell's wishes Ship 2 had two side-by-side seats, so Bell could ride along in his helicopter. Ship 2 had a closed cockpit, whereas Ship 1 was an open cockpit design with only a small windshield. Ship 1 had included no clutch in the drive system, requiring someone to spin the rotor by hand when the engine was started. Ship 2 included a clutch in the power train to facilitate engine starting. Ship 1's controls had differed radically from any preceding, or current, helicopter control system. Young's first arrangement combined a cyclic control and throttle operated by the pilot's right hand and a "pump handle lever" combined collective and antitorque control operated by the left. There were no controls operated by the pilot's feet. Because Floyd Carlson insisted on foot pedals to control the antitorque rotor, Ship 2 and succeeding Bell helicopters were given traditional pilot's controls.

Fortunately Ship 2 came along quickly, because a crash in September 1943 almost completely demolished Ship 1. Carlson flew Ship 1 to the Gardenville Airport to conduct the first autorotational landings of Ship 1. He made the first two without incident, although with a fairly high forward speed. On his third attempt Carlson steeply flared the helicopter to quickly reduce forward speed, sharply bringing the helicopter into a nose-high attitude just before reaching the ground. The helicopter's tail wheel, which was mounted fairly far back on the tailboom, abruptly contacted the ground before the main wheels, bending the tailboom up into the main rotor. Ship 1 then rolled over on its side, and in helicopter pilot parlance, "beat itself to death." Luckily, Carlson escaped uninjured. After sifting through the wreckage, the team members decided that Ship 1 could be rebuilt. All testing for some time to come, however, had to be completed with Ship 2. The delays postponed Larry Bell's first flight in a helicopter until late 1943 or early 1944.

The Gardenville team quickly completed Ship 2 and resumed the test program. Bell invited a number of senior military and government officials to view the helicopter, including top researchers of the National Advisory Committee for Aeronautics (NACA), forerunner of NASA.

Although Bell Aircraft attempted to conduct its helicopter program in secrecy, all of the activity around both the Buffalo, New York, and Gardenville, New Jersey, facilities intensified public curiosity about the strange flying machine. Finally, Bell decided to publicly announce the helicopter program and invited local newspaper reporters to witness a demonstration flight. In May 1944, Carlson flew Ship 2 indoors at the 65th Regiment Armory during a meeting of the Civil Air Patrol in Buffalo, New York. Carlson's demonstration was the first indoor flight in the United States by any type of aircraft. Almost dazzled by spotlights shining in his face and dust kicked up from the floor by the rotor downwash, Carlson never-theless successfully completed several maneuvers inside the armory. The highlight of the demonstration came when Carlson hovered the helicopter with one wheel resting on Arthur Young's outstretched hand.

On July 4, 1944, the original Model 30, modified and rebuilt after its crash, performed before more than 4,000 spectators at a defense workers' rally at the Buffalo Civic Stadium. After Bell's helicopter became public knowledge, the craft performed several rescue and mercy missions. On January 5, 1945, Jack Woolams, a Bell Aircraft test pilot, bailed out of an early model of the Bell P-59, the first U.S. jet fighter. He injured his shoulder but, after slogging through more than a mile of heavy snow, reached a farmhouse near Lockport, New York. Snowdrifts blocked all roads into the region, so Carlson picked up Dr. Thomas C. Marriott in Ship 2 and, in about five minutes, landed him at the isolated farmhouse. The doctor promptly treated Woolams's injuries and frostbitten feet, saving them from amputa-tion. On the night of March 14, 1945, Carlson, again in Ship 2, saved two fishermen who were stranded on an ice floe in Lake Erie for twenty-one hours. After the Coast Guard requested assistance, Carlson conducted some impromptu testing. He had to ensure that someone could safely climb aboard a hovering helicopter. No one at Bell was sure what would happen when someone grabbed the heli-copter. With the assistance of Crew Chief Harry Finagan, Carlson assured himself that he could control the hovering helicopter while someone climbed aboard. Carlson flew about 5 miles out into the lake and rescued the fishermen, one at a time. The rescued men be-came angry when Carlson refused to let them bring along the large string of fish they had caught.

In 1945 the Gardenville team made a decision that would affect the helicopter industry for years to come—a much more momentous decision than they realized at the time. Together they concluded to build a third Model 30, although the agreement between Bell and Young provided that Bell Aircraft would pay for only two helicopters. The team had created so much technology and collected so much data from the previous helicopters that they felt obligated to build a third, incorporating all the lessons learned from Ships 1 and 2. By mounting two motion-picture cameras on the previous helicopters, the team had recorded considerable amounts of performance data that could be reviewed again and again. As a result of the team's de-cision, in January 1945 work quietly began on Model 30, Ship 3, which was as different from Ship 2 as Ship 2 had been from Ship 1.

The engineers retained, however, the two-place enclosed cockpit design. Because they had no funding for a third helicopter, the Gardenville team scrounged materials from other projects to construct Ship 3. On April 20, 1945, the newly designed helicopter took to the air for the first time. When the Bell management learned of Ship 3's success they approved the new helicopter as a flying testbed to provide additional data to the Bell engineering department, which had been assigned to design the first helicopter that would be marketed by the company.

Ship 3's improved performance greatly impressed both Bell's test pilots and passengers alike. The helicopter's superb autorotational capabilities made engine-off landings almost routine, and Bell pilots soon demonstrated them to all passengers. Ship 3 rapidly proved that it was not only an advanced testbed but also very nearly a helicopter that could be marketed commercially. With only a few relatively minor changes in configuration, Ship 3 became the first successful commercial helicopter. Ship 3 suited Bell's plans perfectly. In September 1943 he officially notified the Bell Aircraft board of directors that the company had a prototype helicopter successfully flying and that the company planned to enter the postwar helicopter market.

STANLEY HILLER, JR.

Another young, independent designer, Stanley Hiller, Jr., started his first business, building gasoline-powered model cars at the age of sixteen. When the United States entered World War II, Hiller converted his plant to producing window frames for C-47 cargo planes. That lucrative endeavor, along with his father's assistance, provided the initial funds for his helicopter project. Hiller's father, himself a noted aviation pioneer, encouraged the young man and taught him to fly. In 1937, Hiller saw films of the Fa-61 in flight and determined to build a helicopter of his own. In 1942, still a teenager, he formed the Hiller Aircraft Company and gathered a small group of engineers to build and design his first helicopter.

Hiller decided on a coaxial rotor system. Frenchmen Louis Breguet and Rene Dorand had proven the concept on their Breguet-Dorand gryoplane in 1935 and, although complicated, Hiller believed it to be the best system for his helicopter. Hiller also felt that the outriggers on the Fa-61 added too much weight and that he could not compete with Sikorsky's design. Besides, he wanted to build a helicopter adaptive to civilian commuter use, and the tailrotor was dangerous. In December 1942 his firm began constructing a steel-frame, fabric-covered fuselage that measured 13 feet, 4 inches in length.

As was the case for other inventors, wartime shortages delayed Hiller's progress. His men scrounged parts from every conceivable source and fabricated what they could not find. Hiller could not purchase a reliable power source on the open market, so the 18-year-old convinced Grover Loening, chief aircraft consultant to the War Production Board, to provide a Franklin 90-horsepower engine, derated to 65 horsepower. Hiller later upgraded the engine to a 125-horsepower Lycoming. Workers installed the Franklin engine in 1943 and began ground tests. The first time they ran the engine in their little workshop the engine blast and rotorwash blew out all the skylights. After a cleanup, flight testing began a few days later.

Like most other helicopter designers, Hiller tethered the XH-44 to the ground for the first flights: "X" for experimental, "H" for Hiller, and "44" for the year in which it flew. He had no experience piloting a rotorcraft and was concerned about safety, but his precautions did not preclude accidents. On the first flight, conducted from the family driveway, the tethers were improperly adjusted and the 1,244-pound helicopter tipped over, slightly damaging the handcrafted machine. Hiller shifted subsequent flight tests to the football stadium at the University of California at Berkeley. On July 4, 1944, Hiller flew the bright yellow machine untethered for the first time. On August 30, he flew his helicopter in a public demonstration at San Francisco. Hiller's success so impressed Henry Kaiser, the wealthy builder of Liberty ships, that Kaiser provided enough funding for Hiller to redesign his rotor system.

Hiller introduced not only the first successful coaxial rotor helicopter in the United States but also the first successful all-metal rigid rotorblades. Twenty-five feet in diameter, the metal blades would not cone excessively under aerodynamic stresses, minimizing the danger that they would clash together in flight, thus making the blades safer than wooden blades. His control system consisted of a cyclic control stick, a collective lever, and directional pedals. Unlike the yaw control on a single-rotor helicopter, which incorporated an antitorque tailrotor, the XH-44 pedals alternately feathered the counter-rotating main rotor blades for directional control. Hiller modified his rotor system somewhat to allow the blades to "teeter" in flight, much like the design that Young introduced with Bell Helicopter.

Hiller designed another improved coaxial helicopter but met with no success. His company achieved fame after World War II, however, with an innovative control system for single-rotor helicopters. Contrary to aviation industry forecasters, thousands of pilots did not exit the military and create a boom in personal commuter aircraft. In late 1945, Hiller shifted his focus to commercial and military helicopters. His breakthrough came with the Rotormatic main rotor design. Hiller linked the cyclic pitch controls to a set of small auxiliary blades set at 90 degrees to the main rotor blades. These auxiliary paddle blades dampened deflections in pitch and roll, augmenting stabilization of the helicopter, especially at a hover.

CHARLES KAMAN

In December 1945, with a $2,000 investment, Charles Kaman founded the Kaman Aircraft Company and begin design work on a synchropter, a helicopter with two side-by-side intermeshing rotor blades. Kaman followed Flettner's principles of design but improved the concept with the invention of the servo-flap controlled rotorblades. Kaman organized his company to enter the burgeoning postwar helicopter market and, with his creative genius, recorded several firsts and set many records with his rotorcraft.

The Fa-61 and VS-300 were thus the first practical helicopters, in the sense that they accomplished maneuvers that we now take for granted. Vertical takeoff and landing, hovering, and forward, backward, and sideways flight paved the way for production aircraft that could carry useful loads, thereby performing a myriad of military missions and civilian chores. Although limited in lifting capacity, endurance, and airspeed, the few helicopters that participated in combat service proved the versatility of the new machines, especially in lifesaving and reconnaissance missions. The general manager of Sikorsky Aircraft, Lee S. Johnson, summed up the VS-300's contribution twenty years after its first flight when he said: "Before Igor Sikorsky flew the VS-300, there was no helicopter industry; after he flew it, there was."

REFERENCES

Ahnstrom, D. N. *The Complete Book of Helicopters.* New York: World Publishing Company, 1971.

Apostolo, Giorgio. *The Illustrated Encyclopedia of Helicopters.* New York: Bonanza Books, 1984.

Bilstein, Roger E. *Flight in America.* Rev. ed. Baltimore, MD: Johns Hopkins University Press, 1994.

Brian, Marshall. "How Helicopters Work," http://www.howstuffworks.com/helicopter.htm. Accessed September 2002.

Brooks, Peter W. *Cierva Autogiros: The Development of Rotary-Wing Flight.* Washington, DC: Smithsonian Institution Press, 1988.

Brown, David A. *The Bell Helicopter Textron Story: Changing the Way the World Flies.* Arlington, TX: Aerofax, 1995.

Carey, Keith. *The Helicopter.* Blue Ridge Summit, PA: Tab Books, 1986.

Cowin, Hugh W. *Military Helicopters.* New York: Gallery Books; imprint of W. H. Smith Publishers, 1984.

Delear, Frank J. *Igor Sikorsky: His Three Careers in Aviation.* New York: Dodd, Mead, & Company, 1969.

Dowling, John. *RAF Helicopters: The First 20 Years.* London: Her Majesty's Stationery Office, 1992.

Fredriksen, John C. *Warbirds: An Illustrated Guide to U.S. Military Aircraft, 1915–2000.* Santa Barbara, CA: ABC-CLIO, 1999.

Gablehouse, Charles. *Helicopters and Autogiros; A History of Rotating-wing and V/STOL Aviation.* Philadelphia: J. B. Lippincott Company, 1969.

Glines, C. V. "The Skyhook." *Air Force Magazine* 71 (July 1988).

Gregory, H. F. *Anything a Horse Can Do: The Story of the Helicopter.* Cornwall, NY: Cornwall Press, 1944.

Halcomb, Mal. "The Development History of the Helicopter in Germany from WWI to the End of WWII." *Airpower,* Vol. 37. (March 1990).

Hasskarl, Robert A., Jr. "Early Military Use of Rotary-Wing Aircraft." *Air Power Historian* 12, no. 3 (July 1965): 75–77.

Heatley, Michael. *The Illustrated History of Helicopters.* New York: Bison Books, 1985.

Helicopter's History Site, http://www.helis.com. Accessed June 2002.

Higham, Robin, John T. Greenwood, and Von Hardesty. *Russian Aviation and Air Power in the Twentieth Century.* London: Frank Cass, 1998.

Hunt, William E. *Helicopter: Pioneering with Igor Sikorsky.* Shrewsbury, England: Airlife Publishing, 1998.

Igor Sikorsky Historical Archives, http://www.iconn.net/igor/indexlnk.html. Accessed October 2002.

Keogan, Joseph. *The Igor I. Sikorsky Aircraft Legacy: The Chronology of Fixed-Winged and Rotary-Wing Aircraft of Igor I. Sikorsky and the Sikorsky Aircraft Company.* Stratford, CT: Igor I. Sikorsky Historical Archives, 2003.

Leishman, J. Gordon. "Evolution of Helicopter Flight," http://www.flight100.org/history/helicopter.html. Accessed September 2002.

Liberatore, E. K. *Helicopters before Helicopters.* Malabar, FL: Krieger Publishing Company, 1998.

Pember, Harry. *Seventy-five Years of Aviation Firsts.* Stratford, CT: Sikorsky Historical Archives, 1998.

"Pitcairn PCA-1A." National Air and Space Museum, http://www.nasm.edu/nasm/aircraft/pitcairn_pca.htm. Accessed August 2002.

Ross, Frank, Jr. *Flying Windmills.* New York: Lothrop, Lee, & Shepard Co., 1953.

Saunders, George H. *Dynamics of Helicopter Flight.* New York: John Wiley and Sons, 1975.

Sikorsky, Igor I. *The Story of the Winged-S.* Rev. ed. New York: Dodd, Mead & Company, 1967.

Smith, J. Richard. *Focke-Wulf: An Aircraft Album No. 7.* New York: Arco Publishing Co., 1973.

Smith, J. R., and Antony Kay. *German Aircraft of the Second World War.* London: Nautical and Aviation Publishing Company, 1972.

Spenser, Jay P. *Vertical Challenge: The Hiller Aircraft Story.* Seattle: University of Washington Press, 1992.

Townson, George. *Autogiro: The Story of the Windmill Plane.* Fallbrook, CA: Aero Publishers, 1985.

CHAPTER 3

Korea and the Cold War, 1946–1961

DURING WORLD WAR II a limited number of helicopters from several countries proved the rotorcraft's versatility by performing a number of missions, under hazardous conditions, in several theaters of operation. Helicopter pilots flew both land and sea reconnaissance; directed artillery and air strikes; ferried troops and supplies to remote locations; and, most important, saved the lives of many military and civilian personnel by conducting amazing rescues under previously inconceivable circumstances. A few dedicated pilots, designers, and engineers opened the way for today's global military and civilian helicopter industry.

As post-war military contracts disappeared worldwide, helicopter designers and engineers searched for any innovative manner by which to utilize the unique capabilities of the new aircraft. Unlike these visionaries, most people considered the helicopter an interesting contraption but of little value in civil transportation or other commercial ventures. When military pilots visited the Sikorsky plant to offer their thanks for his wonderful new machine, Igor Sikorsky sought their recommendations on how to improve the performance and commercial success of his machines. Larry Bell envisioned the air ambulance and crop spraying as major roles for civilian helicopters. Leaders of the Soviet Union, both military and civilian, realized the many benefits of operating helicopters in their vast country. Helicopters could go where vehicles and ships could not. Vertical flight machines could resupply isolated villages and military garrisons during the long Siberian winters. Helicopters could also traverse land made so soggy by spring thaws that trucks bogged to their axles. The British experience with helicopters during the war, although limited, convinced leaders of the United Kingdom that they must produce their own helicopters.

BELL HELICOPTERS

Manufacturers and aeronautical engineers spent long hours determining which type of rotor systems might be most effective, and whether large or small helicopters might be the more successful commercially. In early 1946, Bell looked out on his flight line and spied Arthur Young, who designed the semirigid rotor system, test flying a small coaxial helicopter of his own design. Young's new machine incorporated both his and Stanley Hiller, Jr.'s, rotor concepts. Young flew his new helicopter only about ten hours before putting it aside and concentrating on developing the helicopter that became the Bell Model 47.

Following World War II, U.S. helicopter producers envisioned a boom in private helicopter ownership, "an everyman helicopter." Frank Piasecki had demonstrated his version of this type of aircraft by flying his PV-2 from a suburban driveway, refueling at a local filling station, and landing on a nearby golf course for a round of golf. Other designers imagined small helicopters filling suburban garages just as automobiles did. To fill the skies with these commuter helicopters, manufacturers had to receive commercial certification from the U.S. Civil Aviation Administration (CAA) in order to sell helicopters to the public. Two major aircraft manufacturers, Bell and Sikorsky, vied to be the first to certify their helicopters for general private use. Unfortunately for the aircraft companies, federal regulations became so restrictive that helicopter producers abandoned the idea of "an everyman helicopter." Instead they concentrated on several commercial applications, such as air ambulances, crop dusters, mail and passenger carriers, and craft for fighting forest fires.

In January 1945, over Young's objections, Bell moved the entire Gardenville, New Jersey, operation to Niagara Falls, New York. Bell's sagacious decision gave Young's team more floor space, and the helicopter operation soon expanded to more than 100 engineers and workers. The Niagara plant also held the facilities to mass produce Bell's dream of the world's first helicopter with civil certification.

Rumors abounded that Sikorsky planned to submit a civilian version of the R-5 for civil certification, and the competitive Bell wanted to be the first to receive such certification. Several Bell employees later asserted that Bell needed quick certification because he publicly announced that his company would produce 500 CAA-certified helicopters in order to lower the purchase price.

Bell Aircraft engineers and designers at the New York plant were convinced that they held the secret of designing successful aircraft by simply taking production drawings to the assembly plant and producing the aircraft. Constructing successful helicopters, however, proved much more difficult than converting design sheets into a finished aircraft. Bell engineers at the Niagara Falls plant, with no helicopter experience, haughtily ignored Young's Gardenville team and began design work on the Bell Model 42 helicopter. Convinced that the public wanted a helicopter that resembled an automobile, the Bell 42 featured automobile-style doors, leather seats, ashtrays, and an abundance of chrome trim. Bartram "Bart" Kelly, one of Young's pilots, climbed into the Model 42, pulled up on the collective control lever, and was astonished to see the rotor angle pitch down instead of up. Young's crew took the best capabilities of Ship 2 and Ship 3 and incorporated them into what became the Bell Model 47. Bell Aircraft, for a time, had two competing designs (Brown 1995, 40).

The Model 47 proved immanently more successful than the Model 42, and the Bell company dropped the Model 42 after all three prototypes either crashed or were severely damaged in accidents. In addition to the best qualities of Ships 1 and 2, the 47 incorporated irreversible controls to eliminate rotor feedback and a newly designed "bubble" canopy that became the hallmark of the Bell 47. The heat-formed canopy provided great visibility and actually improved the performance of the helicopter. Following the advice of a marketing manager—who told him that people would buy more helicopters if rotorcraft resembled automobiles—Bell insisted that the tailboom framework should be fabric covered. Sales representatives soon discovered that the first thing new helicopter owners did was to cut the fabric off the tailboom framework, and the company adopted the open design. The first twenty-eight models delivered to the U.S. Army for test and evaluation in December 1946, however, had the enclosed tailboom. The U.S. Navy received ten of these aircraft for test and evaluation as well. The Army designated the helicopters the YR-13 and the Navy the HTL-1.

In addition to receiving a contract from the Army and Navy, Bell Helicopter found 1946 an auspicious year in other ways. On March

8 the CAA awarded the Bell Model 47 Type Certification No. 1, the world's first civil certification, allowing Bell to produce the aircraft for civilian sale. On April 5, Young applied for a patent for his helicopter control system, which was granted on May 30, 1950. Young, no longer interested in developing future helicopters, passed the patent on to Bell.

Obtaining the type certification involved several of the Bell employees and government officials. Young, Kelly, Joe Mashman, Floyd Carlson, and several others formed a team to assist the CAA members in establishing new specifications for certifying rotorcraft for private use. Raymond Malloy, the CAA chief engineering test pilot, led the government group in validating the new requirements and regulations for flying civilian helicopters. After Carlson taught Malloy the basics of rotary-wing flight, the combined teams faced a major task in modifying fixed-wing rules in order that helicopters could comply with established government safety and operational standards. Through cooperation, and a very congenial attitude, the certification process required only three months, which was providential for Bell. He had previously ordered 500 Franklin engines and other long lead time components necessary for producing the Bell 47.

After certification by the CAA, Bell initiated a training program for civilians at his Niagara plant. Bell also trained U.S. Army pilots at the Niagara plant until the Army established its own training facility at Fort Sill, Oklahoma. During the certification procedure both Mashman and Carlson received helicopter instructor pilot ratings, which allowed Bell to open a civilian flight school, headed by the newly hired Joe Dunn. A few months later, the company added a training program for civilian helicopter mechanics. The company also ran a flight school for training other military pilots to fly Bell helicopters.

Young's team redesigned the unsuccessful Model 42 and developed several military prototypes from the seemingly unsuccessful project. The first, the Model 48, which the Army designated the R-12, produced an order for two prototypes of the five-place helicopter for test and evaluation. The Army then ordered a production run of thirty-four, but canceled the order in 1947 before any R-12s were actually built. One of the prototype R-12s did, however, set a world's speed record for helicopters of 133.9 mph. The Army soon ordered eleven of a larger version of the Model 48, with the designation of R-12B. This helicopter carried two crew and eight passengers. After taking delivery of the initial test helicopters, the Army ordered no more of that version.

During the late 1940s, Bell Aircraft also introduced the Model

54, which was more akin to the Model 47 than to the Model 42. Bell entered the aircraft in an Army competition for a new liaison helicopter. Despite a second-place finish, the Army ordered three Bell Model 54s, which it designated the R-15; it employed them in evaluations until about 1950.

Although the Model 42 and its successors failed to duplicate the large orders for the Model 47, they laid the foundation for future successes. Bell's engineers profited from designing these larger and more powerful machines. The valuable lessons learned became evident in only a few years, when the U.S. Army sought a new troop-carrying helicopter. Kelly noted that the later success of the UH-1 "Huey" "tracks directly from these early efforts" to design larger machines (Brown 1995, 46).

Between 1946 and 1950, Bell introduced several variants of the Model 47. On December 31, 1946, Bell delivered the first commercial version of the Model 47B, which sported a four-wheel, shock-absorbing landing gear, or optional nylon floats. From 1946 to 1947, Bell sold all seventy-eight 47Bs produced to various civil operators around the United States, most of whom employed them in agricultural work. Mashman had previously demonstrated the adaptability of the helicopter by attaching an improvised spray rig to the Model 30 and dusting the Niagara airstrip with flour. Bell employees filmed the event, proving that the helicopter rotorwash covered the underside of plant leaves with chemical agents, something that fixed-wing crop dusting failed to do. In early 1947, following the agricultural success of the 47B, Bell replaced the original version with a specifically designed agricultural variant. The 47-B3 featured an open cockpit and an extended spray bar, and it carried up to 400 pounds of pesticide dust.

Bell quickly followed with the Model 47D, placing it into production in early 1948. The 47D reverted to the bubble canopy of the original Model 47s. Some early Model 47Ds featured a split bubble canopy, in which the upper part of the bubble and the doors could be removed and replaced by a small windshield. Beginning with the "D" model, Bell produced all 47s with an open-frame tailboom and a four-wheel landing gear until the early 1950s. Capable of carrying three passengers, or 500 pounds of payload, the Model 47D also became the early standard among light helicopters, forming the basis for the H-13B and subsequent military variants. In 1948 the U.S. Army ordered sixty-five H-13Bs, powered by a 200-horsepower Franklin engine, the last Army purchase before the outbreak of the Korean War.

In 1948 the helicopter entered politics as well. Texas congressman Lyndon B. Johnson sought a means to travel the Lone Star State and bring attention to his senatorial campaign. Johnson concluded an agreement with Bell to provide a 47D and a pilot to fly him around Texas. Assigned as the pilot, Mashman attached a loudspeaker system to the helicopter in order that Johnson could fly over small towns and speak to the local citizenry. The 47D performed admirably in the hot, dusty conditions of the Texas summer, proving that the helicopter was more rugged and reliable than many detractors believed (Mashman 1992, 46–68).

Several other achievements proved Bell's and Young's commitment to their helicopter. On May 13, 1949, a Bell 47 set a new altitude record by climbing to 18,550 feet. During the same month another Bell helicopter set a world speed record of 133.9 mph. On September 21, 1950, a Bell 47 became the first helicopter to fly over the Alps. On September 17, 1952, Bell pilot Elton J. Smith set a world distance record when he flew a Bell 47 nonstop from Hurst, Texas, to Buffalo, New York, a distance of 1,217 miles.

HILLER HELICOPTERS

Stanley Hiller believed that his little XH-44 coaxial twin-rotor helicopter would be widely accepted as a home-based commuter, but the market never materialized. Hiller's success, nevertheless, gained the young designer so much notoriety that Bell allowed him to tour Bell's Niagara production facility and had Mashman fly Hiller around in a Model 47. Hiller also purchased a complete Model 47 rotor system. After Hiller departed, Bell told Mashman that what the helicopter business needed was competition, and the new Hiller 360 single-rotor helicopter certainly provided that.

In 1946, backed by funding from Henry Kaiser, Hiller formed United Helicopters and entered the commercial helicopter market. Beginning in July, Hiller, making use of his coaxial rotor concept, introduced a prototype of a design called the "Commuter." The company produced eight of these J-10/UH-4 machines and evaluated them in several tests as a private helicopter. The next year Hiller introduced an updated version, the JUH-5, but neither of the models met with any appreciable success.

In November 1947, United Helicopters introduced the company's first successful helicopter and entered real competition with Bell for the light helicopter market. In fact, because of slow helicopter sales,

Bell had several of the Franklin engines in storage and sold some of them to Hiller when the Korean War expanded military helicopter sales in 1950.

The Hiller Model 360, equipped with Hiller's "Rotormatic" rotor system, held several advantages over Bell's Model 47. The 360 carried three persons, instead of the Model 47's two. The "paddle blade" rotor control system created a very stabilized flight platform, especially at a hover. In some instances Hiller also used more advanced manufacturing techniques than Bell. On October 14, 1948, the CAA issued Certificate No. 6-H-1 for the Hiller Model 360, and in January 1949 the 360 became the first civilian helicopter to fly across the United States. On May 8, 1950, the U.S. Army and Navy issued purchase orders for several Model 360s. The Navy designated its versions HTE-1 and HTE-2. In September the Army ordered an improved model, designating it the OH-23 Raven. The Army used these aircraft in a variety of roles including liaison, command and control, medical evacuation, and as trainers for future pilots. By the time production of the 360 had ended, more than 2,000 of these rotorcraft cruised the world's skies.

In the 1950s, Hiller conducted research into powering a helicopter with ramjets at the tips of a two-bladed main rotor. Hiller produced a total of seventeen HJ-1 and HJ-2 Hornets, fourteen of which the U.S. Army accepted for evaluation as the YH-32 in 1956. A skin of fiberglass and plastic laminate covered the tubular metal frame of the two-place helicopter. Because it produced no torque, the little aircraft had no tailrotor but a conventional airplane tail, canted to counter the rotorwash.

Designed with simplicity in mind, with only a cyclic and collective levers, the Hornet exhibited several shortcomings that limited its versatility. The ramjets proved too loud and underpowered, and their exhaust plume was far too visible at night for military operations. Airspeed was a meager 69 mph, and the engines gulped fuel at such a rate that endurance was about twenty-five minutes at best. The ramjets at the blade tips induced such extreme drag that only a very experienced pilot could expect to land safely. Hiller test pilot Bruce Jones, an experienced World War II and former Bell Helicopter demonstration pilot, felt lucky to be alive after demonstrating an autorotation to Hiller executives (Spencer 1998).

Hiller's success with the Model 360 forced Bell to respond with a lighter version of the Model 47D. The Model 47D-1 weighed 1,000 pounds less than the standard 47 and seated three persons. The decreased weight improved performance and payload.

The U.S. Army later ordered a version of the Model 47D-1, designating the helicopter the H-13E. This version incorporated an upgraded engine, an improved transmission and gearbox, and several structural modifications. The Army eventually purchased more than 490 of the H-13Es, which gained everlasting fame as the medevac helicopters of the Korean War. On November 22, 1950, the U.S. Army's 2d Helicopter Detachment, equipped with H-13s, arrived in Korea and immediately began evacuating wounded to the 8055 Mobile Army Surgical Hospital (MASH).

Charles Kaman continued his experiments with intermeshing rotor designs after World War II. He had flown and improved Anton Flettner's concept of side-by-side intermingling rotors to build several record-setting helicopters. Kaman's synchronized intermeshing rotor design required less power, especially at a hover, because his helicopters had no tailrotor to drain power from the lifting rotors. Although Flettner had produced the first operational military helicopter, Sikorsky's R-4 was the first produced in quantity, and Kaman sought the same commercial success after World War II. On January 15, 1947, he introduced his first commercial helicopter, the K-125, but the aviation community received it with little enthusiasm.

In July 1949, with a flight at Bradley Field, Connecticut, Kaman introduced the improved K-225, which met with more success. Kaman equipped the K-225 with a more powerful 220-horsepower Lycoming engine mounted in an uncovered framework of welded steel tubes that supported tandem seats installed in an open cockpit ahead of the two side-by-side rotor masts. The machine had a small windshield, but Kaman insisted on an open cockpit—no one remembered why. The K-225 was outfitted for night flight, and Kaman probably sought maximum visibility for the crew. The design, nevertheless, left the crew vulnerable to inclement weather. Kaman positioned servo flaps on the trailing edge of the main rotor blades. These flaps, linked to the pilot's cyclic control, deflected the rotor blades up or down, providing directional control. This control system reduced the strength required to fly the helicopter, thus alleviating pilot fatigue on long flights. A vertical stabilizer at the rear of the fuselage increased directional control during forward flight. The aircraft rested on a tricycle landing gear, cushioned by oleo shock absorbers on each gear leg. The tremendous drag of the tandem rotor system reduced the maximum speed of the helicopter to only 70 miles per hour. The U.S. Navy, nevertheless, ordered two and the Coast Guard one, designating them the H-22.

Concerned about satisfying his few customers, Kaman offered to lease the K-225 for crop dusting, thereby eliminating the maintenance costs to the operator. The angled downwash from the tandem rotors dispersed the chemicals in a larger pattern than a fixed wing aircraft, or the Bell 47. Unfortunately few operators knew on which crops to concentrate, what chemicals to use, or the correct application procedure. As a result the K-225 generated few orders. Kaman managed to sell one to the government of Turkey, the first helicopter to fly in that country.

Facing bankruptcy, Kaman convinced the U.S. Navy of the viability of an enclosed cockpit version of the 225, designated the HOK-1. In July 1949 the Navy ordered four prototypes of the HOK-1, and in September 1950 they ordered a training variant of the HOK-1 with the designation of HTK-1. By 1953, Kaman had produced twenty-nine HTK-1s, some serving as utility aircraft during the Korean War rather than in a training role. Despite a more powerful 240-horsepower piston engine, the cockpit modifications reduced the HTK-1's performance, and because it lacked a tailrotor students failed to experience the quirks of the type of helicopters they were more likely to fly after graduation from flight school. Thus Kaman received no more orders for the HTK-1.

Notwithstanding sluggish sales, Kaman's K-225 established two benchmarks in helicopter history. In March 1950, on a delivery flight to the Patuxent River Naval Air Test Center, William R. Murray, a company test pilot, successfully performed the first intentional loop in a helicopter. In December 1951, Kaman installed a 175-horsepower Boeing 502-2 gas turbine engine in the same K-225, producing the world's first turbine-powered rotorcraft. On December 11 the modified aircraft lifted off on its initial test flight, ushering the helicopter industry into the turbine age. Even though the turbine produced less horsepower, its dramatically lighter weight and more efficient performance, especially at higher altitudes, validated Kaman's innovation. The U.S. Navy expressed some interest in the turbine-powered K-225 but placed no orders.

In late 1958 the Kaman Corporation introduced the twin-engined H-43 Husky, a derivative of the K-225. Adopted by the USAF as a short-range air rescue and fire-fighting helicopter, the Air Force began to deploy the Husky worldwide in December 1958. The Husky, initially powered by—and upgraded in the "B" model with two—Lycoming T53-L-1A 860-horsepower turbines, was typical of Kaman's twin intermeshing rotor designs. With a 25-foot all-metal fuselage and a rotor diameter of 51 feet, 6 inches, the small H-43 grossed out at 8,800 pounds. The H-43 regularly carried a pilot, copilot, two firefighters, and a medical technician at a maximum airspeed of 104 knots. When deployed to Southeast Asia in 1964, however, heat and high-density altitudes often reduced the crew to only a pilot and a single parajumper to operate the rescue hoist for downed aircrew rescue operations in the mountainous jungles. Kaman produced 276 of all versions of the HH-43, some of which were used by commercial operators as fire-fighting and crop-dusting machines. Kaman also exported H-43s to Burma, Pakistan, Colombia, Morocco, and Thailand. The USAF withdrew the Husky from service in 1974.

In 1956, the U.S. Navy held a design competition for a compact, high-performance, all-weather multipurpose helicopter. The Navy selected Kaman Corporation's K-20 as the winner, and the next year they awarded a contract to the company for four YHU2K-1 prototypes and twelve production UH-2A helicopters. The Navy also selected Kaman's machine as its interim Light Airborne Multipurpose System (LAMPS) antisubmarine warfare (ASW) aircraft. Kaman designed the Seasprite, unlike his previous machines, as a conventional helicopter, with a 44-foot four-bladed main rotor and a four-bladed tailrotor. First flown in 1959 and powered by a single GE T-58 gas turbine engine, the Seasprite became operational in 1962. The Navy ordered Kaman to retrofit all operational Seasprites, manufactured in transport UH-2, ASW SH-2, and rescue HH-2 versions, with two GE T58-GE-8F 1,350-horsepower turbines, giving the H-2 a maximum takeoff weight of 13,500 pounds, a range of 411 nautical miles, and a maximum airspeed of 130 knots. By 1967 all H-2s were either converted or manufactured as HH-2C or HH-2D. The Navy operated almost 200 Seasprites, with armed variants of the H-2C stationed aboard several destroyers. The Navy armed its H-2Cs with Mk-44/46 homing torpedoes, sonobuoys, and Magnetic Anomaly Detectors for ASW operations. Australia, New Zealand, Egypt, and Taiwan also purchased versions of the Seasprite for their navies, including the SH-2G Super Seasprite.

PIASECKI HELICOPTER CORPORATION

In 1946, Frank Piasecki changed his company's name to Piasecki Helicopter Corporation and gained fame by producing several successful large, powerful tandem-rotored helicopters. Piasecki's unique design allowed the front and rear rotor blades to flap up and down without intermingling and causing a catastrophic failure. The

HRP-1 "Rescuer" proved its worth in search and rescue (SAR), plane guard, and general transportation duties. Instead of rescuing one person at a time like other helicopters, Piasecki's could pick up four or five at once. The HRP-1 also carried a rescue jumper who could enter the water and assist a weak or injured person into the sling of the rescue hoist. The Navy also began developing a dipping sonar for ASW utilizing the HRP-1.

Initially a fabric-covered aircraft, Piasecki began producing the PV-18, an all metal version of the HRP-1. The U.S. Navy initially ordered five of these aircraft, designating them the HRP-2 and in 1950 transferred the helicopters to the U.S. Marine Corps. Unfortunately the new fuselage and side-by-side cockpit configuration added too much weight, reducing the helicopter's useful load, and the five HRP-2s went to the USCG. With a more powerful engine and redesigned rotor hubs the helicopter later reappeared as the PV-22 and achieved great success as the H-21.

In May 1949, eight fabric-covered HRP-1s of Marine helicopter squadron HMX-1 participated in a demonstration at Quantico, Virginia, that would forever alter military tactics. In front of President Harry S. Truman and other high-ranking officials, eight HUP-1s transported forty-two combat-equipped Marines, weapons, and supplies from a simulated aircraft carrier deck to a landing zone located right in front of the spectators. Protected by fighter planes, the dark blue HUP-1s dropped their loads and, within seconds, lifted off again to make the round trip to the carrier for additional loads. This performance aptly demonstrated the new USMC concept of vertical envelopment, upon which the Army later based its doctrine of air mobility.

During the latter days of World War II the U.S. Navy published specifications for a utility helicopter capable of SAR and general transport duties, but small enough to operate aboard aircraft carriers and other large warships. Piasecki's proposed helicopter received the company designation PV-14 and the prototype XHJP-1 by the Navy. In 1948 work began on three preproduction aircraft for Navy evaluation trials, the designation changing to HUP-1 on the production models. In a 1949 fly-off with the Sikorsky S-53, the U.S. Navy determined that the tandem-rotored Piasecki helicopter, which included folding blades, better suited the Navy's future requirements. From 1950 to 1952, Piasecki delivered thirty-two HUP-1 Retrievers to the Navy.

The HUP-1 differed little from the XHJP-1, with the exception of inward-sloping fins attached to the horizontal stabilizers mounted below the rear rotorhead. Shorter than the HRP-1, the HUP-1's 32-foot, 10-inch fuselage consisting of a tubular steel framework covered by thin stressed aluminum accommodated four to five passengers, or three litters, in addition to the two-man crew. For rescue and ship resupply—called vertical replenishment by the U.S. Navy—designers installed a large rectangular hatch in the floor of the forward right side section of the cargo compartment, through which an internal winch could lift up to 450 pounds. Like the HRP-1, the HUP carried a rescue jumper to leap into the water and assist downed pilots into the rescue harness; the increased capacity of the winch could lift both rescued and rescuer, eliminating the need to launch a second helicopter to retrieve the rescue swimmer. A single, centrally mounted 525-horsepower Continental R-975-34 piston engine powered the machine, and both 35-foot three-bladed rotors folded for shipboard stowage. Additionally, for shipboard operations, Piasecki fitted the HUP-1 with a particularly strong, fixed tricycle landing gear. The HUP-1 achieved a maximum airspeed of 119 mph and a range of 275 miles. Most versions included all-weather instrumentation, and several were equipped with ASW dipping sonar.

Tests with a Sperry autopilot allowed Piasecki to eliminate the rear stabilizers and install the more powerful Continental R-975-46 550-horsepower engine in the improved HUP-2. The Navy ordered 165 HUP-2s, with 13 going to the Marines. France bought fifteen for the *Aeronavale;* a few of both the French and U.S. Navy versions were produced as HUP-2s with ASW dunking sonar. The Piasecki Corporation equipped one HUP-2 with a sealed, watertight hull and outrigger twin floats for waterborne tests.

In 1951 the U.S. Army ordered a version of the HUP-2 with a reinforced cargo cabin floor, hydraulically boosted controls, and doors modified to facilitate loading and unloading of stretchers for a general support and medical evacuation helicopter. Beginning in 1953 the Army began taking delivery of seventy of these aircraft, the H-25A, dubbed the "Army Mule." The H-25 proved unsatisfactory for frontline service with the Army, and fifty were transferred to the Navy as HUP-3s, as were three to the Royal Canadian Navy for air ambulance and light cargo duties. The remaining Army models served as trainers and medical evacuation aircraft. In July 1962 a triservice agreement changed the few remaining HUP-2s and HUP-3s in service to the UH-25B and UH-25C. The Piasecki Corporation produced a total of 339 "HUP-type" helicopters, the last in July 1954. The French and Canadian HUP types remained in service until 1966. During the Korean War, Piasecki helicopters received their "baptism of fire." The tandem-rotored helicopters performed transport, liaison, and rescue missions, both on land and at sea. At one time all U.S. aircraft carriers carried at least one HUP-type helicopter for air/sea rescue.

In 1949 the USAF began a search for a helicopter to perform cold weather SAR and to service the Arctic Distance Early Warning (DEW) line. The Piasecki Corporation modified the all-metal PV-18 for the competition, and, on April 11, 1952, Len LaVassar and Marty Johnson lifted off in the YH-21 Workhorse. Weighing more than twice as much as the HRP-2, the craft had a single Curtis-Wright R1820-103 "Cyclone" 1,150-horsepower piston engine that drove the 44-foot three-bladed, counter-rotating, fully articulated tandem rotor system. The Workhorse retained the "Flying Banana" shape, but with a larger 52-foot, 7-inch fuselage and rotor system. Designed to carry up to fourteen combat-loaded troops, or 2,500 pounds of internal cargo, the multimission Workhorse could be equipped with wheels, skis, or doughnut-shaped floats around its wheels for landings on the spongy tundra. The USAF ordered thirty-eight of the aircraft, most assigned to units in Alaska or Canada.

In 1954 the U.S. Army began taking deliveries of the helicopter as the CH-21B Shawnee, with an uprated R-1820-103 1,425-horsepower engine and rotor blades extended by 6 inches. With a maximum gross weight of 15,200 pounds, the CH-21B assault helicopter carried twenty-two fully equipped troops, or twelve stretchers plus two medical attendants. Factory equipment also included an external cargo hook capable of carrying a 4,000-pound slingload. The Shawnee cruised at 85 knots and had a maximum airspeed of 110 knots. The helicopter's manuals listed a service ceiling of 19,200 feet and a range of 400 statute miles with auxiliary fuel tanks. The USAF bought a total of 160 H-21B/Cs and the Army 334. In the early 1960s, U.S. Army crews in Vietnam armed their CH-21B/Cs with 7.62-mm or .50-caliber doorguns. Foreign sales of the H-21 included the West German Army, 26 (built under license by *Weser Flugzeugbau*); French Army, 98; French Navy, 10; Japanese Self Defense Force, 10; and Swedish Navy, 11. French forces operating in Algeria also armed several of their H-21s with machine guns.

Commercial operators ordered a few civil models of the H-21, the Vertol 44, for passenger shuttles and logging operations. New York Airways bought a few fifteen-passenger 44s with baggage compartments, soundproofing, and equipped with large inflatable flotation bags under the nose and rear fuselage to fly routes linking Manhattan/New York City and regional airports. Sabena, in addition to its

Sikorsky S-58s, stationed two Vertol 44s at Brussels heliport. Rick Helicopters in Canada operated ten 44s in logging and general utility work. Piasecki also delivered a single Model 44 (N74056) to Russia.

The Army also funded the Piasecki Corporation's development of the XH-21D (the Piasecki Corporation became the Vertol Aircraft Corporation in March 1966). In this configuration engineers replaced a H-21C single engine with two General Electric T58 turbines. Despite successful flight tests from 1957 to 1958, the variant was not placed in production. In 1962 the H-21B and H-21C received redesignations as the CH-21B and CH-21C. During its service the Shawnee acquired a rather distinctive place in aviation history. A Shawnee dubbed "Amblin' Annie" made the first nonstop helicopter flight coast to coast in the United States. To assist the flight a DeHaviland U-1A Otter refueled the helicopter in flight. The CH-21 became the first U.S. helicopter deployed to South Vietnam in significant numbers, and it gained the dubious distinction of being the first aircraft shot down in which U.S. soldiers were killed: in July 1962 four Army aviators died when their Shawnee was shot down near the Laotian-Vietnamese border. By 1966 the U.S. military had withdrawn most of the CH-21s in service, and by 1969 foreign operators did the same.

On October 23, 1953, Harold Peterson and Phil Camerano lifted off in the Piasecki PV-15, first designated the XH-27, then the YH-16, the largest helicopter flown to that point. Designed as a long-range SAR helicopter to rescue downed USAF strategic bomber crews, the Transporter measured 78 feet in length and featured two three-bladed tandem main rotors, constructed of aluminum skin over a honeycomb aluminum core and measuring 82 feet in diameter. Air Force requirements demanded a range of at least 1,400 miles, necessitating a large cargo compartment to carry internal auxiliary fuel tanks. By installing the two Pratt & Whitney R-2180 1,650-horsepower radial engines and all dynamic components in the upper fuselage, Piasecki designers created a capacious, unobstructed cargo area that, according to the company, would accommodate up to forty troops or three light trucks loaded through the rear cargo doors. This capability drew the interest of the U.S. Army, which joined the USAF in funding further development of the Transporter.

Flight tests indicated that the YH-16 was underpowered. To solve this problem Piasecki engineers replaced the radial engines with two lighter and more powerful 1,800-horsepower Allison T38 turbines, creating the world's first twin-turbine helicopter. In July 1955, Harold Peterson and George Callaghan flew the redesignated YH-16A (company designation P-45) to an unofficial speed record of 165.8 mph. Unfortunately, in December 1956 test instrumentation caused the rear rotorshaft to separate, resulting in a catastrophic desynchronization of the fore and aft rotor blades. The resulting crash caused the cancellation of the project, as well as the planned sixty-nine-passenger YH-16B, powered by two Allison T56-2, 100-horsepower turboshafts. The YH-16B featured innovative interchangeable pods for the rapid transport of such diversified loads as tactical operation centers, field operating rooms, and mobile repair facilities. Piasecki designed a special telescoping, stiltlike landing gear that allowed the YH-16B to accommodate pods of varying sizes. Although canceled, the YH-16 provided the fundamental designs for successful helicopters later produced by both the Vertol and Sikorsky helicopter companies.

In 1946, Sikorsky concentrated on developing a commercial version of the R-5, which generated multitudinous orders from home and abroad. The S-51 seated three passengers in addition to the pilot. The U.S. Navy, Air Force, Coast Guard, and several civilian companies ordered variants of the helicopter. During a demonstration to the U.S. Navy aboard the aircraft carrier *Franklin D. Roosevelt,* Sikorsky test pilot Dimitry "Jimmy" Viner conducted the first naval helicopter rescue by recovering downed Navy pilot Lt. Robert Shields. Sikorsky Aircraft produced a total of 232 S-51s. In 1947 the U.S. Navy ordered three larger versions of the S-51, designated the S-53 by Sikorsky and the XHJS-1 by the Navy. The S-53, designed for SAR, had a 49-foot, three-bladed rotor system powered by a 525-horsepower Continental piston engine; it carried three passengers, but only three prototypes were built (Pember 1998, 35).

Like other manufacturers during the postwar period, Sikorsky suffered a paucity of military contracts, but because of the S-51 and subsequent models, his company experienced growth in commercial markets. On October 1, 1947, Los Angeles Airways, led by Clarence M. Belinn, initiated the world's first helicopter airmail service making use of five S-51s. The airway extended its routes and began carrying passengers in 1954. By the 1960s the airline linked more than 300 communities with Los Angeles International Airport with a route structure covering over 5,000 square miles. In 1947, British European Airways utilized three S-51s to open experimental passenger service in England. In the early 1950s, Sabina Belgium World

Airlines bought several Sikorsky helicopters to open the world's first international helicopter airline, linking Belgium, Germany, The Netherlands, Luxembourg, and France. New York Airways began passenger operations in 1953, linking downtown with La Guardia Airport. Chicago Helicopter Airways began passenger operations in November 1956 and, by 1960, carried more than 300,000 passengers in Sikorsky helicopters. During the late 1950s and early 1960s, regularly scheduled passenger flights by helicopter expanded worldwide.

Several commercial ventures that eventually amassed fleets of helicopters also began operations in the late 1940s, globally expanding during the 1950s and 1960s. By 1965 more than 500 companies employed helicopters in the United States and Canada alone. Okanagan Helicopters, Ltd., in Canada used S-51s to fly in some of the most inhospitable terrain and weather in the world. These companies employed helicopters to search out oil and mineral deposits and to service drilling sites and pipelines from the Arctic Circle to the Gulf of Mexico. Humble Oil Company (Exxon) initiated a program of ferrying work crews to offshore drilling platforms with helicopters. Helicopters enabled crews on offshore rigs to spend more time with their families, and sick or injured workers could be flown directly to hospitals. Other companies operating helicopters soon followed in conducting geological surveys; building defense sites in remote locations; constructing power transmission lines; conducting timber and wildlife surveys; and fighting forest fires. The use of helicopters saved both time and money, with the savings sometimes paying for the cost of a helicopter in just a few months of operation. Military operators observed civil operations and learned methods of reducing maintenance and parts replacement costs by increasing time between overhauls.

Sikorsky did not possess a monopoly on the oil company use of helicopters. Bell pilots "Tex" Johnson and Dick Stansbury demonstrated the adaptability of the Bell 47 by flying geologists and a gravity meter over the coastal marshes near Galveston, Texas. As a result of these demonstrations, Bell soon formed Bell Helicopter Supply Corporation to fulfill contracts with oil companies to operate three helicopters in the Louisiana swamps. Bell pilots transported seismographic gear and personnel in a three-year mapping of the oil-rich Louisiana bayou country. Bell 47s also ferried crews and essential tools to work sites, both on and off shore. One of the best known U.S. helicopter companies used the Bell 47, and Bell name, to great advantage. In the late 1940s several oil company executives formed

Petroleum Bell Helicopters, but soon afterward changed to Petroleum Helicopters Incorporated (PHI).

In the late 1940s, Sikorsky attempted to generate more military interest with several new designs, but these proved unfruitful. Sikorsky installed his first set of all-metal rotor blades on the S-52, but the diminutive two-place machine generated only slight interest among military officers. The S-52 did, however, set a couple of altitude and speed records, and in a later test it became the first Sikorsky helicopter to be looped. Sikorsky installed a gas turbine engine in an S-52, designating the aircraft the S-52-5, the company's first turbine helicopter. The S-53 produced only the three-ship Navy order. The company modified an R-4 to produce the S-54, a tandem-rotored test project, soon abandoned.

THE S-55 HELICOPTER

The company's next venture, the S-55, however, garnered international acclaim and produced large orders from several countries. Internally funded by Sikorsky Aircraft and under development for only seven months, the S-55/YH-19 lifted off on an initial flight on November 10, 1949, and hit the market the next year. As the USAF and U.S. Army H-19 and U.S. Navy and Marine Corps HO4S, the S-55 saw combat service in both Korea and Vietnam. Both the British and French military also purchased the H-19.

An all-metal pod and boom design, two pilots sat high in a cockpit situated above and behind the rounded forward engine compartment, which housed a single Pratt & Whitney R-1340 550-horsepower radial engine. With the H-19, Sikorsky and other designers began separating the crew, engine, and cargo compartments. Sikorsky installed his standard three-bladed main rotor, but with all-metal blades designed to withstand the exigencies of combat flying for hundreds of hours. A two-bladed tailrotor projected from a gearbox near the end of a long, slim tailboom extending from the oval-shaped fuselage. The centrally located boxlike cargo compartment accommodated more than 2,600 pounds of cargo, nine to twelve passengers or troops, or up to eight stretchers and medical supplies when modified as an air ambulance. For the first time medical personnel rode on board to care for the wounded in transit to hospitals. On production models Sikorsky replaced the horizontal stabilizer on the right side of the tailboom with two anhedral tail surfaces and added a triangular fin between the underside of the tailboom and fuselage. Situated below the floor of the cargo compartment, two crash-resistant fuel tanks gave the helicopter a range of more than 250 miles. Most military versions had a four-wheeled landing gear, but some U.S. Navy and Air Force models were manufactured with permanent metal pontoons that functioned as amphibious landing gear.

The USAF ordered a total of 270 H-19Bs with a 700-horsepower Wright R-1300-3 engine and a larger main rotor. In 1952 the U.S. Army ordered 72 of the H-19C Chickasaw, the first of which arrived in Korea in January 1953 with the 6th Transportation Company. On April 28, 1950, the U.S. Navy ordered the first of 119 S-55s, the last delivered in January 1958. The Navy's various versions included 10 HO4S-1s (equivalent to the H-19A) and 61 HO4S-2s (about 30 of these were built as the HO4S-3G for the USCG). The Marines received 99 HRS-2s and 84 HRS-3s, which corresponded to HO4S, designed as troop transports. Redesignated the UH-19 in 1962, several variants served both the U.S. Army and Marines into the mid-1960s.

By July 1953, H-19 crews flew all manner of missions in Korea. H-19s rescued UN aircrews from behind enemy lines; ferried troops and supplies to critical points of combat; conducted reconnaissance missions; recovered damaged vehicles; evacuated casualties from the battlefield; and hauled almost any conceivable load. By the time of the cease-fire they were carrying more than 70 tons of cargo a month to UN forces operating in South Korea. One of the reasons for the success of the H-19 was ease of maintenance. Sikorsky designed the helicopter for ease of access to parts and assemblies. The engine, for example, could be changed in two hours without the necessity for special tools.

The S-55 claimed several firsts for helicopters, being the first used in antisubmarine warfare. Two H-19s assigned to the U.S. Air Force Rescue Service, named *Hop-a-long* and *Whirl-a-way*, completed the first transatlantic crossing by helicopter between July 15 and July 31, 1952. The Air Force crews flew the 3,410 miles from Westover Air Force Base in Massachusetts to Germany by way of Labrador, Greenland, Iceland, and Scotland in 42.5 hours of flight time. Along the way, the crews set another record by flying a 940-mile leg of the route nonstop. The S-55 received the first type certification from the CAA as a transport helicopter on March 25, 1952. In 1952 the S-55 also became the first rotary-wing craft used for commercial airline links in Europe. The Belgian airline Sabena flew

S-55s between the chief cities in Belgium, Lille, Rotterdam, Bonn, and Cologne. Sikorsky produced a total of 1,281 S-55 models in the United States, and another 477 were built under contract in England, France, and by Mitsubishi Heavy Industries in Japan.

Certainly part of the demand for the S-55 may be directly attributed to the success of the Sikorsky H-5 during combat operations in Korea. On July 22, 1950, the first of nine H-5s assigned to the USAF 3rd Air Rescue Squadron arrived at Taegu and almost immediately began evacuating wounded soldiers from the battlefield. The H-5 proved so effective that, on August 14, the commander of the Far Eastern Air Force, General George E. Straterneyer, requested twenty-five additional H-5s for the medical evacuation role. On September 4, 1950, a USAF H-5, covered by a Combat Air Patrol, conducted the first rescue of a downed airman behind communist lines during the Korea War. During the Korean War, Air Rescue Service helicopters ferried 9,216 injured military personnel, of those 8,598 from front line units, to Mobile Army Surgical Hospitals (MASH). U.S. Navy HO3 "Dragonflys," dubbed "Angels" by carrier pilots, plucked several downed Navy aviators from the freezing waters off the Korean peninsula.

Sikorsky H-5 helicopters also evacuated UN soldiers trapped behind enemy lines. On October 22–23, 1950, two H-5s and three L-5 liaison aircraft rescued 47 paratroopers surrounded on drop zones at Sunchon and Sukchon, deep in enemy territory near Pyongyang. On March 24 and 25, 1951, a single Sikorsky YH-19 helicopter joined H-5s to evacuate 148 paratroopers cut off on the Munsan-ni drop zone north of Seoul.

As the Korean front stagnated in 1951 the Air Rescue Service established bases on Paengnyong-do and Cho-do islands located off the west coast of Korea, and conducted several dramatic helicopter rescues from those islands. During World War II a pilot downed behind enemy lines was virtually assured of capture or worse, but helicopters increased the chance of recovery during the Korean War. Between June 1950 and the July 1953 cease-fire, the Air Rescue Service plucked 1,690 USAF aircrews from enemy territory, 170 by helicopters. Additionally, another 84 airmen from other services and UN allies were rescued from behind enemy lines.

Several operational limitations, however, plagued the H-5. Lack of armor, limited range, and the capability of lifting only four personnel, including the pilot, made missions behind enemy lines extremely precarious. An early loss during the war aptly demonstrated the vulnerability of the H-5. On October 1, 1950, an H-5 returning to Allied lines along a road detonated an enemy antitank mine with its rotorwash. The explosion blew the light rotorcraft into a hillside.

The longer range, increased cargo capacity, and better altitude and speed capabilities of the H-19 made the helicopter a welcome addition to the UN arsenal. In February 1952, H-19s began replacing all H-5s in the rescue inventory.

Aviation companies known for fixed-wing designs also experimented with helicopters during the years following World War II. James S. McDonnell introduced the XHJD-1 Whirlaway on April 27, 1946. Equipped with two Pratt & Whitney R-985 450-horsepower engines, the machine was designed by McDonnell with the U.S. Navy in mind. Unfortunately the aircraft did not impress Navy officials, and only one unit was built. On May 5, 1947, McDonnell produced the world's first ramjet helicopter. Powered with ramjets mounted at the rotor tips, the XH-20 Little Henry had no tailrotor and weighed only 280 pounds. The project was canceled because of excessive fuel consumption. A proposed H-29 improved version was also canceled.

In 1952, Cessna Aircraft Company executives postulated that a light helicopter would be an asset to the company's range of light fixed-wing aircraft. They also understood that Cessna's L-19 Birddog light reconnaissance aircraft might soon be replaced by a helicopter and wanted to protect the company's military sales. To fill the expected niche in the marketplace, Cessna bought the Seibel Helicopter Company. Charles M. Seibel, an experienced helicopter engineer, left Bell Helicopter and established his own company and developed a series of experimental helicopters in the late 1940s. Seibel delivered two S.4 SkyHawk two-place helicopters to the U.S. Army for evaluation. The S.4, of conventional design, was powered by a 108-horsepower Lycoming 0-235-C1 piston engine. The machine, designated YH-24 by the Army, failed to generate any interest. Seibel then built the S.4A three-seater and the two-seat S.4B with a 165-horsepower engine, but neither found any commercial success.

Seibel's technological expertise nonetheless impressed Cessna officials. Cessna bought Seibel's company and directed him to design a high-performance two-seat personal helicopter. He created the CH-1. Calculated to resemble Cessna's single-engined aircraft, it made use of an all-metal monocoque fuselage with a conventional empennage, and a streamlined Plexiglas cockpit with 360-degree vision for the pilot. He powered the machine with a supercharged 260-horsepower Continental engine, mounted in the nose and connected to the main rotorshaft by a flexible driveshaft. After a series of test flights the CH-1 received its type certification in June 1955, and later that year it earned some notoriety by landing on Pike's Peak (14,000 feet).

With little or no market for a two-seat helicopter, Cessna executives decided to expand the seating to a four-place aircraft within the confines of the existing CH-1 airframe. Desiring to market the helicopter to the U.S. military, Cessna secured an order for ten YH-41 Senecas for evaluation, and the first was delivered in December 1957. The YH-41 proved both underpowered and unstable, as well as exhibiting poor autorotational characteristics; that ended any hope of a military contract. Cessna eventually bought back six aircraft, converting them to civilian CH-1Cs.

In 1960, Cessna introduced the CH-1C Skyhook, with increased power and a modified control system that somewhat resolved the controllability problems. Incomplete and sketchy records indicated that the company produced forty-five Skyhooks, with exports to Argentina and Canada. Additionally, Iran bought five and Ecuador six, under the Military Assistance Program. In 1963, because of a series of engine and transmission malfunctions, Cessna ceased all production of the CH-1.

WESTLAND HELICOPTERS

Across the Atlantic Ocean, European companies began manufacturing helicopters of their own. After their exposure to the unique capabilities of rotorcraft during World War II, the British decided to invest in helicopters. A successful aircraft manufacturer since 1915, Westland Ltd. decided to concentrate on producing helicopters. In 1946 the company entered into a license agreement to produce the Sikorsky S-51, with the designation of WS-51. After modifying the helicopter to meet British airworthiness standards, Westland nicknamed their version the "Dragonfly." On October 5, 1948, Westland rolled out the first of 149 versions of this helicopter. On August 23, 1955, Westland introduced the Widgeon, an updated version of the Dragonfly. Westland enlarged the cabin to accommodate a pilot and four passengers, added metal rotor blades, and increased performance with an Alvis Leonides 521/1 520-horsepower engine. Range increased to 270 nautical miles.

Another British company, Bristol Helicopters, founded in 1944, designed and built a multipurpose helicopter during the late 1940s, under the leadership of the helicopter pioneer Raoul Hafner. On July 27, 1947, the Bristol Type 173 Sycamore took to the air for the first time. The helicopter featured a long, slim fuselage, cradling a 550-horsepower piston engine. The rotor blades measured 45 feet in diameter. Carrying a pilot and four passengers, the rotorcraft reached a maximum airspeed of 114 knots and a range of 270 nautical miles. The Sycamore measured 39 feet in length and had a rotor diameter of 44 feet, 11 inches. The company produced a total of 183 models of this aircraft, most of which were sold to the RAF and RN for transport and SAR. The Federal Republic of Germany (FRG-West Germany), however, did buy a few machines for its Air Force and Navy.

Bristol Helicopters produced one other helicopter before the company folded. The Type 192 Belvedere was a larger, turbine-powered tandem rotor design that appeared in July 1958. The Belvedere, designed as a general transport helicopter, looked very similar to the Piasecki machines, with two three-bladed rotors at each end of the fuselage. The Type 192 joined the RAF in 1961 but, although powered by two 1,300-horsepower Napier Gazelle NGa2 turboshafts, did not perform as well as expected. The Bristol Company found it difficult to compete with Sikorsky, Bell, and Westland, and it produced only twenty-six machines. In March 1960, Westland absorbed all interests of Bristol Helicopters.

Another company later assimilated into Westland also designed a successful helicopter in the late 1940s. Saunders-Roe, which acquired the Cierva Autogyro Company on January 22, 1951, tested its Skeeter in October 1948. The Skeeter carried a pilot and one passenger/observer. A Gypsy 215-horsepower inline piston engine drove a three-bladed wood and canvas rotor measuring 32 feet in diameter. With a 28-foot fuselage, the helicopter's maximum takeoff weight was 2,190 pounds. It reached an airspeed of 87 knots and had a maximum range of approximately 190 nautical miles. The little helicopter served as a trainer with both the RAF and West German Air Force. All the companies involved produced a total of eighty-eight Skeeters.

After Westland had absorbed all the major UK helicopter production in 1960, in addition to producing helicopters under license, the company began to build helicopters of its own design. The first, the re-engineered Saunders-Roe P-531, which became the Westland Mk I Scout, rolled out on August 4, 1960. A single Rolls-Royce Nimbus 101 1,050-horsepower turboshaft, derated to 710 horsepower, powered the five-place general-purpose helicopter. The

Scout measured 40 feet, 3 inches in length and the main rotor 30 feet, 9 inches in diameter. With a maximum gross weight of 5,300 pounds, the helicopter attained a maximum airspeed of 105 knots and had a range of 270 nautical miles. Westland built a total of 149 Scouts, most of which served with the British Army in training and observation roles.

Across the English Channel, just prior to World War II, France nationalized its aviation industry, resulting in the formation of the Societe Nationale de Constructions Aeronautiques Sud-Est (SNCASE) and the Societe Nationale de Constructions Aeronautiques Sud-Ouest (SNCASO). After the war, both companies, subsidized by the French government, initiated a myriad of projects to develop rotorcraft to meet a wide variety of requirements. In early 1947, Sud-Est began flying a rotorcraft, albeit of antiquated design. The SO-3000, a twin-rotor six-seat model, was copied directly from the German Fa 223. As in the Fa 223, cantilever outriggers extended outward and upward from each side of the welded steel-tube fuselage, supporting the two three-bladed rotors. A German Bramo Fafnir BMW 323 R2 1,000-horsepower radial engine powered the French machine. The two-man crew sat in a Plexiglas-covered cockpit. A cabin behind the pilot's position accommodated four passengers, and a small hoist provided power for lifting cargo and rescue work, but the SO 3000's lack of success generated little interest and fewer orders.

The previous year Sud-Ouest had introduced its first helicopter, the SO 1100 Ariel, also derived from experiments conducted by German aeronautical engineers during World War II. A two-seat, all-metal aircraft, the Ariel featured small tip jets powering the three-bladed main rotor, but again buyers did not appear.

In June 1948, Sud-Est tested a prototype SE 3101, an experimental model to test an unconventional twin-rotor antitorque system. Powered by an 85-horsepower Mathis piston engine, the little one-seater elicited no sales, but it provided the platform for a much more successful machine.

On March 23, 1949, Sud-Est began testing the SE 3110, the progenitor of a successful line of French-designed helicopters. Resembling the Bell 47, with a two-seat cockpit covered with a clear "bubble" canopy, a separate engine compartment immediately behind the crew stations, and a long, tubular steel tailboom, Sud-Est's craft had a 200-horsepower Salmson 9-piston engine, boosted by a Turbomeca compressor. French engineers used an articulated three-bladed main rotor and a conventional two-bladed tailrotor on this helicopter. The landing gear consisted of skids attached to the underside of the open engine compartment.

In 1951, Sud-Est constructed two SE 3120 prototypes, a three-seat improvement on the SE 3010. The company named the new version Alouette (Lark) and also installed the Salmson engine in these prototypes, but with better results. On July 31, 1952, one of these helicopters established a world record by remaining aloft for thirteen hours and fifty-six minutes.

On December 16, 1953, Sud-Ouest test pilots lifted off in the radical SO 1221 Djinn. Developed from the Ariel models, the Djinn provided a modicum of success for the French aeronautical industry. In addition to the French Army, ten countries ordered 178 units of the unusual rotorcraft. A Palouste gas generator powered the tip jets. The craft contained no electronic system, and the engine had to be started by hand with a large crank sticking out of the starboard side of the fuselage. In 1954 the U.S. Army received three Djinn aircraft, designated YHO-1, but demonstrated little interest in the French design. Several of the helicopters went into service as crop sprayers.

Seeking more performance from their aircraft, Sud-Est began testing gas turbine engines in their helicopters. On March 12, 1955, Jean Boulet lifted off in the first of two prototype SE 3130 Alouette IIs, powered by Artouste II 360-horsepower gas turbine engines (Apostolo, however, lists the first flight as 1953). Three months later the Alouette II set a new world altitude record for a light helicopter by climbing to 26,925 feet. Following that success, Sud-Est completed three more prototypes for further testing.

In 1956, prompted by orders for 363 machines from the French Aeronavale (Fleet Air Arm), followed by others from the French Army and Air Force, the SE 3130 became the world's first turbine-powered helicopter to go into full production. Primarily designed for observation, liaison, SAR, and casualty evacuation, the Alouette II, expanded to carry four passengers and a pilot, appeared in armed versions. The French Army equipped some SE 3130s with AS-10/11 antitank missiles or machine guns, and the French Navy adapted a few to carry homing torpedoes for ASW. The United Kingdom, Federal Republic of Germany, and Austria also ordered substantial numbers of the helicopters. When production of all types of the Alouette II ended in 1975, forty-six countries were operating more than 1,300 aircraft.

On May 2, 1957, the Alouette II received a French domestic certification, opening the way for sales to civil operators. The same year the French government merged Sud-Est and Sud-Ouest, forming Sud Aviation; the helicopter's designation changed to SE-313B. Saab in Sweden and Republic Aircraft in the United States received licenses to manufacture and sell the SE 3130. All military and civilian versions could be supplied with skid, wheel, or pontoon landing gear. In 1969, Sud-Est also launched a high-altitude version called the Lama, which was also built under license in Brazil, and by Hindustan in India, and called the Cheetah. On June 21, 1972, a Lama set a world helicopter altitude record of 40,800 feet.

On February 28, 1959, Sud-Aviation tested a SE 3180 Alouette III prototype. Modified with a more powerful 550-horsepower engine and stronger gearboxes and transmission, the SE 3160 (SE 316A, and later SA 318) went into production in 1960. Some twenty versions of this helicopter, the last versions with a Turbomeca Astazou 870-horsepower engine, remained in production until 1975. Like its predecessor, the Alouette III succeeded in both military and civilian roles, especially as a crop sprayer. When production of the aircraft ended in 1983, seventy-four countries operated 1,455 Alouette IIIs.

On June 10, 1959, Sud-Aviation's first heavy-lift helicopter, the SE 3200 Frelon, lifted off for its initial test flight. Embarrassed by its lack of a nationally designed and produced large helicopter, the French government directed that Sud-Aviation produce a helicopter comparable to the Piasecki and Sikorsky transport helicopters in French service. Of conventional design, with a single main and tail-rotor, it had three Turbomeca Turmo III turboshaft engines mounted atop the fuselage. In May 1965 the production model, SA 3210 Super Frelon, appeared in both troop carrying and naval ASW variants, with more powerful Turmo IIIC turbine engines. The production models featured a boat-hulled fuselage with a rear loading ramp and a wheeled landing gear. It was intended to carry a crew of three and twenty combat troops; France manufactured ninety-nine machines and sold several to Israel, Libya, and Iraq. The People's Republic of China produced a licensed version called the *Changhe Z-8*.

NICOLAI I. KAMOV

Although Josef Stalin had imprisoned the majority of the USSR's most promising helicopter engineers in 1935, two remained free to become the country's most prolific helicopter designers. During the great patriotic war, Nicolai I. Kamov had produced both successful autogyros and his first helicopter, the Ka-8 Vertolet. Under Kamov's tutorship Dr. Mikhail Leontyevich Mil studied rotary-wing aerodynamics, created his own original designs, and, in 1947, became head of the Moscow Helicopter Plant.

Born on November 22, 1909, the son of a railroad employee and a dentist, Mil entered the Siberian Technological Institute at Tomsk in 1926. Two years later he transferred to the Don Polytechnical Institute, where he concentrated on aeronautical engineering. In 1929, Mil participated in the design of one of the Soviet Union's first autogyros, the Kaskr. In 1931 he joined the Central Aero and Hydrodynamic Institute (CAHI), where he headed a special aerodynamic experimentation team and published works on autogyros, rotor-wing aerodynamics, and general theories on rotorcraft. In 1940, Mil became deputy chief designer of Helicopter Plant 1290, where he continued his designs of autogyros under the direction of Kamov. Wartime necessity forced the evacuation of the plant, but Mil continued to study the stability of hinged rotor systems, which became the basis of his 1945 Ph.D. dissertation in aerodynamics.

After World War II, Mil returned to CAHI, where he designed an experimental helicopter in 1945. The EG-1 was a three-place, traditional concept with a single main rotor and an antitorque tailrotor. In April 1946 he submitted the design to the Soviet Ministry for Aviation Industry. The ministerial board accepted his proposal, with some adaptations for civilian aircraft. On March 27, 1947, the Soviet government established Aviation Laboratory 15 to perform experiments with rotary-wing flight and appointed Mil as director of the new facility. The personnel from this lab formed the initial engineering team for the Mil Helicopter Design Bureau, established on December 12, 1947. Mil led this bureau and the Moscow Helicopter Plant 1329 until his death on January 31, 1970.

Mil's experimental helicopter first flew in 1947, evolving into the Mi-1 Hare, which lifted off for the first time in September 1948. The Mi-1's pod and boom fuselage sat on a fixed tricycle landing gear. A light alloy skin covered a welded steel tube fuselage and tailboom framework. The three main rotor blades, like most others of the era, were constructed of wood. A single Ivchenko AI-26V 575-horsepower radial engine powered the light helicopter, but Mil almost immediately began experimenting with gas turbines to power his helicopters. His experiments resulted in a modified Mi-1, the Mi-2 Hoplite, in which Mil installed two Isotov GTD-350 400-horsepower turboshaft engines. Despite a number of successful turbine experiments the Mi-2 did not take to the air until 1961. The next year the Soviet Union authorized WSK-Swidnik (now PZL), located just outside Warsaw, exclusive production rights to the Mi-2 Hoplite.

In the late 1940s and early 1950s an experienced Soviet aircraft designer endeavoring to compete with Mil met with scant helicopter success. A. S. Yakolev had produced several successful fixed-wing aircraft during World War II, and he placed his Yak 100 helicopter into competition with the Mi-1. Yakolev's helicopter greatly resembled Sikorsky's S-51, but it apparently lacked the S-51's performance. First flown in 1950, the Yak 100 lost a fly-off to the Mi-1, which was adopted by the Soviet military. Yakolev introduced the Yak 24 in 1953. This helicopter mirrored Piasecki's tandem rotor design. By 1955 the Yak 24 was the largest flying helicopter in the world, lifting 8,000 pounds to an altitude of 6,000 feet. Known in the USSR as the "Flying Boxcar," the Yak 24 received the NATO designation "Horse." The civil version carried thirty passengers, but large Mil helicopters overshadowed the Yak 24.

In addition to military aircraft, the Mil plant produced several civilian variants of the Mi-1 and Mi-2, including light transport, air ambulance, passenger, and agricultural versions. Mil's first designs proved successful in many aspects. During the late 1950s and early 1960s, the Soviet collective farm system used the agricultural versions of these helicopters to spray millions of acres with fertilizer during the late winter and early spring, increasing Soviet crop production by 25 percent.

Soviet Fleet Admiral N. G. Kuznetsov, impressed by Kamov's Ka-8, proposed to the Soviet Aviation Ministry that Kamov be placed at the head of a design bureau specifically tasked to design rotorcraft for the Soviet Navy. On October 7, 1948, as a result of the admiral's influence, the Minister of Aviation Industry formed the Experimental Design Bureau No. 2 (OKB), also known as the Kamov Design Bureau, headed by Kamov himself. The Aviation Ministry housed the OKB in Experimental Factory No. 3 in Sokol'niki, a district of Moscow.

Kamov remained convinced that his coaxial concept provided a more efficient lift coefficient to power ratio, and in September 1949 he introduced the Ka-10. Receiving the NATO designation Hat, the Ka-10 was an improved Ka-8 equipped with a 55-horsepower Ivchenko engine that propelled the Hat to airspeeds of about 50 knots. Kamov debuted the twin vertical fintail stabilizers on a late model of the Ka-10; the twin vertical fins becoming a distinguishing feature of his aircraft. Like Yakolev, Kamov achieved little success with this model, and only twelve machines were built. Several Soviet Navy officers, however, expressed interest in a more powerful Kamov design.

Until 1950 helicopter manufacturers worldwide had been unable to convince military commanders to spend any appreciable portion of their budgets procuring rotary-wing aircraft. Most military officers had little concept of the helicopter's adaptability, viewing rotary-wing aircraft as mere odd-looking gadgets. An Army general, before the outbreak of the Korean War, told Larry Bell that if he really wanted to make his helicopter useful, he should find a way to power the wheels, "so that after you land, you could drive right up to the field headquarters" (Brown 1995, 100).

When the army of the People's Republic of North Korea invaded the Republic of Korea on June 25, 1950, the concept of helicopter employment changed forever. During the summer and fall of 1950, U.S. helicopters, supporting UN forces in defense of South Korea, assumed a multiplicity of roles, including liaison, reconnaissance, artillery adjustment, medical evacuation, and resupply. During the next three years of combat on the Korean Peninsula, helicopters validated the versatility promised by helicopter designers and engineers, resulting in a boom in sales of both military and civilian rotorcraft.

Bell Helicopter received orders for more machines during one of the company's most tumultuous periods, but it managed to fill all of its contracts. About the same time as the conflict erupted in Korea, Bell decided to move his rotary-wing production facilities. A nineteen-week labor strike in 1949 caused Bell to search for a new plant location. In 1950 he began to transfer all helicopter manufacturing from Buffalo, New York, to Fort Worth, Texas. After the war began the U.S. Army ordered an additional eighty-five H-13Ds, which were delivered in 1951, just prior to the move of all production facilities to Texas. Bell modified these H-13s with recently developed skid landing gear, instead of the normal four wheels. Landing on the steep ridges of Korea was much easier with skids, as wheels tended to roll off the steep terrain. Bell received a patent on skid landing gear on June 9, 1953. These and all subsequent Bell 47s/H-13s appeared without the tailboom covering.

The air ambulance version of the H-13, fitted with a litter attached to each skid, soon acquired the name "Angel of Mercy" from UN combat troops. The H-13 transported the majority of the 21,000 casualties evacuated from the front lines by helicopter during the Korean War. The speed with which a wounded soldier reached field hospitals in Korea reduced the death rate of combat casualties to half that of World War II. The success of the air ambulances resulted in a follow-on purchase of 265 OH-13Gs by the U.S. Army. The "G" model was the first H-13 equipped with all-metal rotor blades. In all, Bell and its licensees produced more than 6,000 Model 47/OH-13s before production ended in 1963.

Beginning in 1952, Kawasaki Heavy Industries, under an agreement with Mitsui & Co., which held authority to build Bell helicopters in Japan, also opened an assembly line for Bell 47s and later Bell products. One of KHI's most successful early designs was the KH-4, an improved version of the Bell Model 47. The KH-4 employed a new control system, increased fuel capacity, and a higher useful load. The KH-4's redesigned cockpit held three passengers on a rear bench seat behind the pilot.

U.S. Marine experimentation with vertical envelopment tactics in 1949 reaped benefits during the Korean War. Natural obstacles on the craggy Korean peninsula posed almost insurmountable problems of moving troops and supplies by traditional methods. The Marines partially solved this problem by ferrying men to virtually inaccessible mountaintops by helicopter. Marine Transport Squadron 161, equipped with fifteen HRS-1s, arrived at Pusan on August 30, 1951, and began transporting troops and supplies to the front lines; it returned to base with wounded Marines and soldiers. On October 11, 1951, Marines conducted the first battalion-size air movement of troops in history. In six hours several HRS-1s, each carrying six combat-equipped Marines, inserted 950 Marines onto a mountaintop landing zone. The success of Marine helicopters in lifting troops to the front lines and evacuating the wounded prompted General Matthew B. Ridgeway, commander of UN forces in Korea, to recommend that four Army helicopter companies be sent to the Far East Command. Ridgeway realized the unique capacity of helicopters to exploit situations by quickly inserting combat forces to critical points, and the advantages that helicopter resupply had over paradrops.

The U.S. Marines also attempted to bring the first armed helicopter into action. Marine Corps officers at Quantico, Virginia, related to Hans Weichsel and Joe Mashman that both the Marines and Army faced a tremendous challenge destroying the Soviet-built T-34 tanks employed by the North Koreans. U.S. Marines and infantrymen employed the bazooka as their major antitank weapon. The shoulder-fired weapon delivered a short-range, antitank rocket, but, because of the rough terrain, bazooka teams often could not maneuver close enough to engage the enemy tanks. Both the Marine officers and the Bell executives agreed that an armed helicopter could be used as a very effective tank killer.

Weichsel and Mashman, through diligent effort and at the cost of a couple of cases of Scotch whiskey, located a bazooka and a few rounds of ammunition. Most of the weapons were en route to Korea. They improvised a mount for the rocket launcher on the skid and rigged a temporary firing mechanism. The sighting system was an aiming dot marked on the canopy, calculated to allow Mashman to hit a target a few hundred yards away. Weichsel and Mashman decided to test fire the weapon while the helicopter sat on the ground. The two men placed balsa wood sticks on the tailboom and tailrotor, and then fired a bazooka round while the helicopter ran at full operating rpm. The back blast of the rocket motor did not harm the balsa strips, so the next event was an airborne firing at a target (Brown 1995, 58; oral history interview, Hans Weichsel, Jr., 2002).

Mashman, with a Marine colonel on board, lifted off the airfield and flew to the Quantico firing range. He hovered the H-13 about 300 yards from a large board with a dot painted in the center. Mashman aligned the crude sights on the target and fired the bazooka. To his astonishment, the rocket hit the target almost dead center. He calmly turned to the colonel and asked, "Does that convince you of its accuracy?" The Marines at Quantico were ecstatic with the success, but Marine Corps headquarters killed the concept of armed helicopters. General officers were afraid that if the United States armed H-13s the North Koreans would have a valid excuse to shoot at medical evacuation helicopters. Most junior officers were aghast at this reasoning, because the North Koreans already fired on any helicopter that flew into range (Weichsel 2002, 58).

Fierce combat on the inhospitable Korean peninsula unconditionally confirmed the obvious. The U.S. military clearly required larger helicopters powerful enough to transport more troops and heavier combat loads, such as vehicles and artillery pieces. Both Sikorsky and Piasecki responded to that need. Piasecki's solution, a re-engineered HUP-2, took flight in April 1952. The all-metal CH-21 retained the "banana" shape and tandem rotor design, but with side-by-side pilot and copilot stations. The Wright R-1820 piston engine produced 1,425 horsepower, enough to lift twenty fully equipped troops, or 4,800 pounds of internal cargo, or twelve litter cases, at a cruise speed of 90 mph. The entire aircraft, including rotor arcs, measured 86 feet, 4 inches in length.

Although originally designed as a cargo and troop transport, the CH-21 served in a variety of roles, with a number of military forces. The "flying banana" carried troops and cargo; rescued downed airmen and sailors; evacuated wounded from the battlefield; and resupplied construction crews at high altitudes and in the Arctic. The helicopter played an important role in the construction of the U.S./Canadian DEW line in Alaska and Canada by hauling construction crews and equipment to locations otherwise unreachable. The USAF initially purchased 18 YH-21s, and followed that order with 38 H-21A Work Horses and 163 H-21Bs with more powerful engines. The U.S. Army ensured the aircraft's success by purchasing 334 H-21C helicopters, called the Shawnee.

On December 18, 1953, Sikorsky's first twin-engined helicopter, the S-56, made its maiden flight. Initially designed to meet a 1950 Marine request for a heavy-lift assault helicopter, the helicopter—christened HR2S-1 by the Navy Department and H-37 Mojave by the Army—received significant funding for its development from the U.S. Army. A 1962 joint service resolution universally designated the Mojave as the CH-37 by all U.S. military forces.

The CH-37 featured a large fuselage with clamshell doors at the nose, capable of holding twenty-six troops and all their equipment, or three jeeps, or a 2.5-ton truck, or light artillery pieces. As an air ambulance, the 1,500-cubic-foot cargo compartment held twenty-four stretchers. A long tailboom extended from the upper portion of the fuselage, with a four-bladed tailrotor placed high on a vertical fin. To achieve the maximum cargo space the CH-37's designers installed two 1,900-horsepower Pratt & Whitney R-2800 Double Wasp piston engines in nacelles at the end of short, shoulder-mounted wings. The nacelles also contained the retractable main landing gear. A 68-foot, five-bladed rotor system, designed to fly with one blade shot away, topped the new machine. Sikorsky Aircraft later increased the rotor diameter on subsequent models to 72 feet and boosted power to 2,100 horsepower per engine. The later models could lift 11,000 pounds of cargo, and all cargo versions were fitted with an internal winch capable of hoisting 2,000 pounds.

In 1954 the Army borrowed a Marine preproduction HR2S-1 for rigorous flight testing. As a result of the excellent performance of the helicopter during that evaluation, the Army ordered ninety-four similar versions, designated the H-37 Mojave. Sikorsky delivered the complete Army order before June 1960. On October 25, 1955, the first of sixty production HR2S aircraft scheduled for the U.S. Marines took flight, and Marine Corps Squadron HMX-1 began receiving its machines the next July. Production of the CH-37 ended in May 1960, but the Sikorsky plant was engaged until the end of 1962 in converting all but four of the helicopters to "standard" models. Sikorsky modified two HR2S with an AN/APS-20E search radar, mounted in a bulbous radar dome under the nose, marking the HR2S-1W patrol helicopter as a Navy Airborne Early Warning aircraft.

Primarily because of the expense of operating a piston engine–driven helicopter of that size, commercial markets for the S-56 did not materialize. Although the new Lycoming T55 gas turbine engines were available in 1955, proposals to install them in the CH-37 were not adopted. The heavy radial engines, however, did not prove a complete disadvantage. Sikorsky built 156 H-37s, which remained the largest Western helicopter for many years. Additionally, the aircraft held two international height-with-payload records from 1956 to 1959.

Sikorsky Aircraft's next major contribution to military helicopter development, the S-58, first took flight on March 8, 1954. Sikorsky produced the helicopter as a result of the U.S. Navy's search for a larger, more powerful craft to replace the HO4S/S-55 in ASW. Called the HSS-1 by the Navy, the first production model lifted off the flight line on September 20, 1954, and became operational in August 1955. The U.S. military redesignated the helicopter, nicknamed the Seabat, the SH-34G in 1962. The original model, and the later variants of the SH-34J and the "winterized" LH-34D, all carried either a dipping sonar or torpedoes with which to attack submarines. The newer versions were outfitted with more sophisticated electronic tracking gear and a flight stabilization system. The H-3 Sea King replaced the Seabats in the late 1960s.

In April 1953 the Army placed preliminary orders for the S-58 as the H-34 Choctaw, and in 1955 the Marines ordered them as the HUS-1 Seahorse, the two almost identical troop-carrying versions. In April 1955 the Army received the first of 437 H-34As and an additional 21 HUS-1 aircraft transferred from the Marine Corps, also designated H-34A.

Capable of carrying twelve combat-equipped troops or 2,700 pounds of internal cargo, the S-58 had a fuselage that measured 54 feet, 9 inches. Similar in design to the S-55, with a forward engine compartment and a flight deck above and slightly forward of a large cargo area, the S-58 featured many improvements, including advanced communications and navigation radios. A single Wright R1820 1,525-horsepower engine drove the 62-foot, four-bladed main and tailrotors. The landing gear consisted of two main wheels and a tail wheel on the tailboom that was thicker than the H-19's and sloped down slightly. Most versions had an external hook capable of transporting a 3,500-pound slingload.

The performance of the H-34 met or exceeded all expectations by the military. For example, in 1956 an early production model, flown by U.S. Army Captains Claude E. Hargett and Ellis Hill, set new world helicopter speed records on courses of 100, 500, and 1,000 kilometers. It was judged the first helicopter safe enough for routine use by the U.S. president, and the Army, in 1957, established an Executive Flight Detachment equipped with specially modified Choctaws. Sikorsky fitted each CH-34 with extensive soundproofing, plush VIP interiors, and advanced avionics. President Dwight D. Eisenhower regularly flew aboard these VH-34As, the first sitting president to travel by helicopter, his first flight being in a Bell Model 47J. In 1960, Sikorsky began modifying U.S. Army and USAF H-34As to a standard model "C" with the addition of automatic flight stabilization systems and other detail changes. By January 1962 the Army had 190 H-34Cs and 179 H-34As in its inventory. The Marines received about 500 S-58s, mostly cargo/utility versions. Under the Tri-Service designation system introduced later that year, the military redesignated the helicopters as, respectively, CH-34C and CH-34B. The Army later upgraded several "C" model aircraft to VH-34C standards. Sikorsky produced 2,261 of the machines, and the Choctaw remained on active duty with the U.S. Army until the late 1960s, and with reserve units much longer. The last Choctaw did not officially retire until the early 1970s.

From the mid-1960s, exported variants of the S-58, and those produced abroad under license, served with a variety of armed forces and civilian airlines. The Federal German Army bought 144, the navies of Argentina and Brazil purchased 5 each, France operated 126, and Belgium 9. Italy's military put 18 into service, Japan 14, Israel and The Netherlands 12. Indonesia, Cambodia, Canada, Thailand, and South Vietnam also bought several dozen of the military versions of the S-58. SUD Aviation manufactured most of the French and Belgian S-58s under license in France.

Sikorsky also supplied several commercial S-58B and S-58D passenger/cargo transport helicopters to civil operators. The CAA certificated the twelve-seat airline version in August 1956, and a sixteen-passenger version in 1970. Chicago Helicopter Airways, New York Airways, and Sabena Airlines all bought versions of the S-58. Although production of the S-58 ended in 1965, Sikorsky offered a civilian conversion of the S-58 powered by two PT6T-6 gas turbine engines in 1970.

In 1956, Westland acquired a license to build the S-58 and developed a turbine-engined version called the Wessex. Initially Westland imported an HSS-1 and flight tested the aircraft with its original Wright engine. The company then modified the machine with a 1,100-horsepower Napier Gazelle gas turbine engine and improved metal rotor blades, designed to sustain more combat damage. On May 17, 1957, the modified rotorcraft made its first flight, and on June 20, 1958, two preproduction Wessex Mk 1s, designed for naval trials, followed. Upgraded with a 1,450-horsepower Gazelle Mk.161 engine and equipped with a dipping sonar and homing torpedoes, the HAS Mk.1 went into production in 1959 for the Royal Navy (RN) as an ASW helicopter. RN No. 700H Flight received the first models in April 1960, and the next year six RN squadrons added the new aircraft to their inventory; the first No. 815 Squadron received theirs in July 1961. The RN equipped 848 Squadron with Wessex designed for commando assaults and stationed them aboard HMS *Albion*. These aircraft carried sixteen fully equipped Royal Marine commandos or eight stretchers and a medical attendant in the cargo area. Optionally, the commando version could deliver a 3,600-pound slingload on an external hook. In January 1967 the HAS Mk.3 entered service. Powered by a 1,600-horsepower Gazelle Mk 122, Westland installed advanced search gear in a large dorsal radome. In August 1962, Westland began supplying the Royal Australian Navy with a Wessex model similar to the HAS Mk 1 but powered by a 1,540-horsepower Gazelle Mk 162 engine. Westland delivered a total of 320 military variants to UK armed forces and another 42 versions to foreign militaries. The Queen's Flight detachment received all-red HCC 4s with VIP interiors and advanced avionics.

On May 24, 1958, the prototype of Sikorsky's first amphibious helicopters, and the first in a long line of successful turbine-powered machines, took to the air. The single-engined S-62, which became the USCG HH-52A Seaguard, actually flew before the twin turbine S-61. Sikorsky designers adapted the H-19's rotor system to a 44-foot, 6.5-inch boat-hulled fuselage with two outrigged floats housing retractable landing gear. The single General Electric CT58-GE-8 1,250-horsepower turbine provided enough power for the Seaguard to carry up to twelve passengers, or six litters, and a crew of three to a maximum speed of 109 mph at sea level and climb to a service ceiling of 11,200 feet. With a maximum fuel load the helicopter reached a range of 474 miles. In addition to a hoist, the engineers included a specially designed folding rescue platform for the helicopter. On January 9, 1963, the USCG accepted the first of 99 HH-52s. The company produced a total of 170 military and commercial versions of the helicopter, the first turbine-powered helicopter certified for commercial use by the FAA. HH-52s stationed aboard icebreakers were painted orange with a white slash on the tail, a paint scheme still utilized by the USCG. USCG historians claimed that, because the Seaguard remained in service until 1985, the HH-52A rescued more persons than any other helicopter in the world. Mitsubishi Heavy Industries produced 25 military and civilian versions of the helicopter, some used for SAR with the Japanese Air Self Defense Force (JASDF).

FRENCH INDOCHINA

About the same time that U.S. forces began utilizing helicopters in Korea, the French military in their war against the communist Vietminh introduced the first two helicopters into French Indochina (Vietnam). Although French companies produced some rotorcraft, the medical service of the French Air Force, Far East, brought two Hiller 360 light helicopters into Indochina in 1950. The 360s carried only one pilot and one passenger and were used for medical evacuation. Unfortunately, the Hillers proved too small and underpowered to operate anywhere except in the lowlands of the Mekong Delta. Larger and more powerful Sikorsky H-5s and H-19s soon augmented the French helicopter force. Only the H-19s, however, produced enough power and lift to operate in the high-density altitudes of the mountains of Vietnam, but even they proved marginal in certain conditions. By the end of 1952 the French had ten helicopters in Indochina and built a heliport on the outskirts of Saigon (Ho Chi Minh City). By the war's end in 1954, forty-two U.S.-manufactured helicopters, mostly Sikorsky, served French forces in Indochina. Those few helicopters built an impressive record in medical evacuation and supplying isolated units. Virtually all received hits by enemy fire, but only two were shot down. French pilots quickly learned to fly at 3,000 feet above the ground, out of machine gun range, and usually followed secure roads to their destinations. Fighter aircraft, when available, escorted helicopters during missions in which French planners expected enemy resistance. During the Indochina War, French helicopters ferried 10,820 sick and injured men to medical facilities and rescued 38 pilots forced to bail out of their aircraft. Helicopters also rescued eighty bloody stragglers from the doomed bastion of Dien Ben Phu in May 1954.

Only six months after their devastating defeat in Indochina, French forces faced a colonial war of independence much closer to home. On November 1, 1954, members of the National Liberation Front (NLF) began terrorist and guerilla operations in Algeria. In 1955 the French military responded with the first of 400,000 troops and operated U.S.-designed helicopters in the mountains and deserts until Algerian independence in July 1962. The French Army and Navy introduced Sikorsky H-19s in 1955 and Piasecki H-21s in 1956. A naval squadron equipped with Sikorsky HSS-1s appeared in 1958. To conduct an all-out mobile war against the separatists, the French modified several of these helicopters to carry machine guns and bombs, and at least one of the HSS-1s had a 20-mm cannon. The armed helicopters proved so effective in counterguerilla operations that U.S. military planners considered arming U.S. helicopters supporting the Army of the Republic of Vietnam in the nascent war against the Vietcong.

The French operated their helicopters (some of the Sikorsky S-58s manufactured under license by SUD Aviation) much the same as in Indochina. The rotorcraft conducted troop insertions/extractions, medical evacuations, SAR, and resupply missions throughout Algeria, especially in the mountain strongholds of the rebels. French Navy Corsairs flew close air support (CAS) for many of the helicopter assaults. French heliborne raids became so successful in 1958 and 1959 that many guerrillas fled into sanctuaries in Morocco and Tunisia. During the Algerian war for independence French helicopters flew more than 49,000 hours, losing at least seven helicopters to combat action and accidents. The losses resulted in the deaths of thirteen aircrewmen.

The helicopters of Aerotecnica in Spain had their roots in research conducted in the 1950s by Jean Cantinieau in France. Cantinieau, an engineer with SUD Ouest (SNCASO), in cooperation with M. Decroze, made use of triangulated tubing to construct an open-frame single-seat machine. The two men mounted the engine and rotor above the pilot's head to reduce mechanical linkages between engine and rotor. The conventional design included a long tubular tailboom and small tailrotor. The C 100 rested on a three-wheeled landing gear. On November 10, 1951, the C 100 made its first of three flights at St. Cyr and performed well enough for Cantinieau to build a two-seat model.

The MC 101 retained the general structure of the C 100 but included a more powerful 105-horsepower engine. On November 11, 1952, Gerard Henry lifted off in the first of two MC 101s. Impressed by the little helicopter, the Spanish industrialist Marquis del Merito, who owned Aerotecnica SA, an aerial photography and crop spraying corporation based near Madrid, decided to bring the little airship to Spain.

In 1953, del Merito funded Cantinieau's move to Spain, where the C 101 became the Aerotecnica AC 11. Madrid's higher altitude and hot climate necessitated more power, and a Lycoming 150-horsepower engine was installed. The Spanish government funded two follow-on AC 12 prototypes with a full, all-metal monocoque fuselage with panels encompassing the engine compartment, which housed a O-320-B2A, 170-horsepower piston engine. Cantinieau installed a three-bladed main rotor above a fully enclosed two-seat cabin and changed the landing gear to skids rather than wheels.

Cantinieau sold a design for a three-seat turbine-powered machine to Societe Nationale de Constructions Aeronautiques de Nord (SNCAN). The French company built two prototypes of the helicopter, an all-metal aircraft with a large bubble canopy and a Turbomeca Artouste I turbine mounted above and behind the cockpit, which turned a three-bladed main rotor. Cantinieau replaced the conventional tailrotor with a futuristic ducted exhaust gas arrangement similar to that employed some twenty-five years later on the Hughes, then McDonnell-Douglas, NOTAR designs. The pilot controlled the ducted exhaust gas, which counteracted main rotor torque, through conventional pedals.

On December 28, 1954, the three-seat Norelfe prototype flew successfully, but SNCAN sold both the aircraft and all rights to Aerotecnica. After further testing the redesignated AC 13A, Aerotecnica expanded the cabin to a five-seat version, the AC 14. To compensate for the additional weight engineers installed a more powerful Turbomeca Artouste I1B 400-horsepower turboshaft.

The Spanish government placed orders for twelve piston-engined AC 12s and ten of the AC 14s. The Spanish Air Force redesignated the aircraft EC-XZ-2 and EC-XZ-4, respectively. The helicopters served for only a relatively short time before they were retired.

Aerotecnica planned several other helicopters, but none came to fruition. The company initiated construction of a prototype of the twelve- to fourteen-passenger AC 21, powered by twin Turbomeca

Turmo turbines and a massive ducted-air tailboom. The company envisioned a turbine version of the AC 12 and an AC 15, actually an AC 14 powered by a Lycoming O-435-V 260-horsepower engine. In 1962, however, the Spanish government terminated all funding to Aerotecnica, and the company folded.

Cantinieau returned to France and joined Matra, where he designed the "Bamby." This single-seat aircraft resembled his earlier designs but incorporated a triangular end to the tailboom, which more efficiently controlled the ducted gases. During testing in 1963 the Bamby experienced several failures, and Matra abandoned the project.

Other designers continued to improve their concepts, which resulted in new models exhibited late in, or shortly after, the Korean War. In the early 1950s the Soviet Navy issued specifications for a helicopter capable of lifting heavy loads, performing a variety of missions, and operating in adverse weather. In 1952, responding to the Navy's request, Kamov rolled out the Ka-15 Hen prototype, in which he installed the newly available Ivchenko 225-horsepower radial engine near the center of the helicopter. The two-seat multipurpose rotorcraft reached an airspeed of 80 knots and a service ceiling of over 10,000 feet. The compact fuselage, with no tailrotor, required only small landing areas and smaller storage space for shipboard operations.

Kamov's improvements to the Ka-10 met the Soviet Navy's requirements, but the new helicopter had to prove its merit. On April 14, 1953, the Ka-15 made its first flight, and the Soviet Navy ordered a fly-off between the single-rotor Mi-1 and the coaxial Ka-15 to determine which helicopter could best operate aboard ship. The fly-off took place aboard the cruiser *Mikhail Kutuzov,* which had a small helipad. The Ka-15's small size and agility made possible several successful landings and takeoffs, even in rough and windy seas. The Mi-1's long tailboom and rotor, however, limited maneuverability, and the helicopter could not safely land or take off when high winds created turbulence over the ship, or when rough seas caused the cruiser to pitch and roll. The results of the fly-off convinced naval officers that the Kamov coaxial design met all of their prerequisites for a helicopter.

Kamov maintained the three-bladed coaxial rotor system but enlarged the Ka-15's fuselage and engine. A pilot and observer/equipment operator sat side by side in the small cockpit. Behind the crew a cargo compartment held the electronic equipment required for specific missions. The ASW version carried two RGB-N sonobuoys or the SPARU-55 automatic airborne receiver unit. One Hen dropped sonobuoys in the water, and another received information from the sonobuoys if they detected a submerged submarine. Once a submarine was detected, a third Ka-15 equipped with the OPB-1R sight and two 50-kg depth charges swooped in to attack. The first units, equipped with all three types of the ASW Ka-15 helicopters, joined the Soviet fleet in 1957. By 1961 the Soviet Navy had launched at least eight ships equipped with helipads and support equipment for Ka-15s and their crews.

Sundry versions of the Ka-15, with some improvements, served various roles for almost twenty years. A civilian version, the Ka-15M, with chemical hoppers filling the cargo area, served mainly in agricultural roles, spreading fertilizer and battling insect pests and weeds. An air ambulance version was fitted with two internal litters. The OKB installed dual controls in the UKA-15 for instructing pilots to fly the counter-rotating coaxial helicopters. In 1958 the OKB began testing composite rotor blades on the Ka-15M. The new blades increased the lift/drag ratio of the rotor system and extended the service life of the blades. In 1958 and 1959 test pilot V. V. Vinitsky established two world speed records with the Ka-15M. In 1960, OKB upgraded most Ka-15Ms with a 275-horsepower AI-14VF engine. The Ka-15 marked the beginning of coaxial helicopters operating with the Soviet Navy and with Aeroflot, the Soviet state airline.

In 1955, OKB introduced the Ka-18 Hog, a Ka-15 with a larger fuselage and a 280-horsepower Ivchenko engine. The updated helicopter accommodated four passengers, or more cargo. OKB produced approximately 200 of these aircraft in both military and civilian versions.

Observing U.S. success with rotorcraft during the Korean War, Mil recognized the tremendous military potential of helicopters for the USSR. In 1952 he introduced the first of his large helicopters. The Mi-4 Hound greatly resembled, but was much larger than, Sikorsky's H-19. Mil designed the Mi-4 with a large, oval-shaped fuselage, equipped with clamshell cargo doors at the rear. He mounted a single Shvetson ASH-82V, 1,700-horsepower radial engine in the nose section. Like the H-19, the cockpit was stepped-up above the engine compartment. The large engine and a four-bladed main rotor system provided enough lift to carry up to sixteen combat-equipped troops or 5,350 pounds of cargo internally. A long, high-mounted tailboom supported a three-bladed antitorque rotor and horizontal stabilizers. The Mi-4, as with many Mil designs, sat on a four-wheeled landing gear. With a cruising speed of 100 mph the Hound had a range of 250 miles. Although it was designed primarily as a cargo and troop transport for the Soviet Army, a 12.7-mm machine gun could be mounted to the fuselage underside, and some versions carried two 57-mm rocket pods attached to a frame on the sides of the fuselage. The Soviet Navy also ordered several Mi-4s equipped for ASW roles. The Mil factory manufactured approximately 3,500 Mi-4s before production ended in 1969.

For a country comparable in size to the United States, and with a much larger military, China operates very few helicopters. The Z-5 (Zhishengji-5), China's first domestically produced helicopter, was a slightly modified version of the Soviet Mil Mi-4 Hound. In February 1958, Harbin Aircraft Factory (now Harbin Aircraft Manufacturing Corporation, HAMC) began producing the Z-5 under an agreement with the USSR, which had agreed to assist the Chinese government in equipping the People's Liberation Army (PLA) with modern weapons systems. The first Chinese-built Z-5 lifted off for the first time in December, with major production scheduled for 1959. Major deficiencies in quality control in the Chinese factories caused the PRC government to refuse issuance of a production certificate until 1963. By the time production of the Z-5 ended in 1980, a total of 545 units had been built for the PLA Air Force (PLAAF) and Chinese Civil Aviation.

A 1,770-horsepower Huosai-7 piston engine powered the Z-5, which gave the helicopter a maximum cargo payload of 1,500 kg. On later variants metal rotor blades replaced the wooden ones of earlier models. Harbin added two external fuel tanks to increase range. To offer increased cargo-handling capabilities the company installed an electric hoist and a hatch in the floor of the cargo compartment. Other modifications increased the maximum speed from 185 km/h to 210 km/h. The last models of the Z-5 could hover in ground effect at 18,000 feet. An armed variant of the helicopter carried a 12.7-mm heavy machine gun mounted on the underside of the fuselage and two 57-mm rocket pods attached to the sides of the helicopter.

CONCLUSIONS

The period between World War II and the early 1960s brought the helicopter into world prominence, both as a modern weapons system and a civilian workhorse. Whether plucking wounded soldiers and Marines from craggy ridgelines or downed aircrewmen from the freezing waters off the coast of the Korean peninsula, the renown earned by U.S. "choppers" substantiated the versatility, and unquestioned practicality, of rotary-wing aircraft. From the frozen vastness of Siberia, to the steaming mountains and jungles of Indochina, to the deserts of North Africa, helicopters performed tasks outside the capabilities of conventional fixed-wing aircraft. Rotorcraft, once considered tangential gadgets, earned esteem from skeptical military and civilian leaders by completing previously unimaginable tasks.

Operating in combat situations and such diverse climatological environments, however, exposed numerous shortcomings of the fledgling helicopter industry. Technology limited helicopter performance in many areas. Heavy reciprocating engines limited available power, thus reducing speed and the maximum gross weights of rotorcraft. The lighter turbine engines, with more power-to-weight efficiency, ameliorated that problem. The turbine-powered Kaman 225 proved this when the helicopter climbed to 10,000 feet and the turbine operated so efficiently that the test pilot had to close the throttle and autorotate to a lower altitude. At high altitudes the helicopter continued to climb at idle power. The turbine also reduced the noise levels in helicopters to just a turbine whine and the noise of the rotor blades. In 1955, Lycoming introduced the T-53 series gas turbine engine designed especially for installation in helicopters, which resulted in machines designed especially for power and speed. Helicopter engineers in Europe and the USSR also made use of the benefits of turboshafts in their helicopters.

Metallurgy also increased the efficiency and durability of helicopters. Revolutions made in manufacturing fixed-wing aircraft during World War II were adapted to helicopters. Helicopters produced with lightweight aluminum and magnesium alloy skins and frames could lift larger loads. Metal rotor blades replaced laminated wooden blades, increasing both lift and resistance to the elements. Lighter airframes also allowed for the installation of self-sealing fuel tanks and armor for the crew and critical engine parts, which had proved extremely vulnerable to groundfire in early machines.

Nascent industries on both sides of the Iron Curtain (NATO and the Warsaw Pact) vied with one another to produce the most efficient and deadly helicopter on earth. Military planners demanded larger and more powerful rotorcraft, capable of surpassing the capabilities of all their adversaries' helicopters. Generals and admirals yearned for helicopters powerful enough to attack an enemy's troops, planes, and ships. Helicopter innovators readily complied with these requests because of the financial rewards and the personal satisfaction involved.

Additionally, helicopters took on new roles in the civilian market. The undisputed success of air ambulances in Korea laid the foundation for the boom in civilian EMS helicopter services in the latter half of the twentieth century. Civilian helicopters, usually modified from military designs, offered passenger service in areas never previously served. Helicopters carried petroleum engineers, drilling and construction crews, and loggers into the most unimaginable sites throughout the world. Helicopters also became television and movie stars. In the 1950s the TV series *Whirlybirds* was based around the Bell 47. Moviemakers also realized the great versatility of the helicopter as a camera platform and used the machines to shoot scenes never before thought possible. Newspaper publishers saw the benefits of sending reporters out in helicopters to "scoop" the competition. The *Chicago Tribune* bought a Bell 47, becoming the first U.S. newspaper to own its own helicopter.

Although no market materialized for the "everyman helicopter," helicopters played a role in the financial recovery after World War II. Military and civilian operators from around the world placed orders with helicopter manufacturers, increasing the companies' profits and providing thousands of additional jobs for both white- and blue-collar workers. Fortunately for both military and civilian operators, helicopter designers and manufacturers looked to the future, which would place even greater demands on their aircraft.

REFERENCES

Ahnstrom, D. N. *The Complete Book of Helicopters.* New York: World Publishing Company, 1971.

Apostolo, Giorgio. *The Illustrated Encyclopedia of Helicopters.* New York: Bonanza Books, 1984.

Bell Helicopter-Textron, http://www.bellhelicopter.textron.com. Accessed June 2002.

Bilstein, Roger E. *Flight in America.* Rev. ed. Baltimore, MD: Johns Hopkins University Press, 1994.

Boeing Company, http://www.boeing.com. Accessed January 2003.

Brian, Marshall. "How Helicopters Work." http://www.howstuffworks.com/helicopter.htm. Accessed September 2002.

British Army Air Corps Museum, http://www.flying-museum.org.uk. Accessed May 2002.

Brown, David A. *The Bell Helicopter Textron Story: Changing the Way the World Flies*. Arlington, TX: Aerofax, 1995.

Carey, Keith. *The Helicopter*. Blue Ridge Summit, PA: Tab Books, 1986.

Chant, Christopher. *Fighting Helicopters of the 20th Century*. Christchurch, Dorset, England: Graham Beehag Books, 1996.

Cowin, Hugh W. *Military Helicopters*. New York: Gallery Books; imprint of W. H. Smith Publishers, 1984.

Delear, Frank J. *Igor Sikorsky: His Three Careers in Aviation*. New York: Dodd, Mead, & Company, 1969.

Dowling, John. *RAF Helicopters: The First 20 Years*. London: Her Majesty's Stationery Office, 1992.

Everett-Heath, John. *Soviet Helicopters*. London: Jane's Publishing Company, 1983.

Fort Rucker, AL. Home of U.S. Army Aviation, http://www.rucker.army.mil. Accessed April 2002.

Gunston, Bill, ed. *The Encyclopedia of Modern Warplanes*. New York: Barnes & Noble Books, 1995.

Harding, Stephen. *US Army Aircraft since 1947*. Stillwater, MN: Specialty Press, 1990.

Heatley, Michael. *The Illustrated History of Helicopters*. New York: Bison Books, 1985.

Helicopter World, http://www.helicopter.virtualave.net. Accessed April 2002.

Helicopter's History Site, http://www.helis.com. Accessed June 2002.

Helicopters of the U.S. Army, http://www.geocities.com/capecanaveral/hangar/3393/army.html. Accessed August 2002.

Higham, Robin, John T. Greenwood, and Von Hardesty. *Russian Aviation and Air Power in the Twentieth Century*. London: Frank Cass, 1998.

Hirschberg, Michael, and David K. Daley. *US and Russian Helicopter Development in the 20th Century*. Np: American Helicopter Society International, 2000.

Igor Sikorsky Historical Archives, http://www.iconn.net/igor/indexlnk.html. Accessed October 2002.

Kelly, Orr. *From a Dark Sky: The Story of the US Air Force Special Operations*. Novato, CA: Presidio Press, 1996.

Keogan, Joseph. *The Igor I. Sikorsky Aircraft Legacy: The Chronology of Fixed-Winged and Rotary-Wing Aircraft of Igor I. Sikorsky and the Sikorsky Aircraft Company*. Stratford, CT: Igor I. Sikorsky Historical Archives, 2003.

Kuznetsov, G. I. *Kamov OKB 50 Years, 1948–1998*. Edinburgh: Polygon Publishing House, Birlinn Publishing, 1999.

Lightbody, Andy, and Joe Poyer. *The Illustrated History of Helicopters*. Lincolnwood, IL: Publications International, 1990.

Mashman, Joe (as told to R. Randall Padfield). *To Fly Like a Bird*. Potomac, MD: Phillips Publishing, 1992.

McGuire, Francis G. *Helicopters 1948–1998: A Contemporary History.* Alexandria, VA: Helicopter Association International, 1998.

Palmer, Norman, and D. Kennedy Floyd, Jr. *Military Helicopters of the World.* Annapolis: Naval Institute Press, 1984.

Pearcy, Arthur. *U.S. Coast Guard Aircraft since 1916.* Shrewsbury, England: Airlife Publications, 1991.

Pember, Harry. *Seventy-five Years of Aviation Firsts.* Stratford, CT: Sikorsky Historical Archives, 1998.

Ripley, Tim. *Jane's Pocket Guide: Modern Military Helicopters.* London: Jane's, 1997.

Royal Air Force Official Site, http://www.raf.mod.uk. Accessed May 2002.

Royal Navy Official Site, http://www.royal-navy.mod.uk. Accessed May 2002.

Russian Aviation Museum, http://www.2.ctrl-c.liu.se/misc/ram. November 2002.

Shrader, Charles R. *The First Helicopter War: Logistics and Mobility in Algeria, 1954–1962.* Westport, CT: Greenwood Publishing, 1999.

Sikorsky, Igor I. *The Story of the Winged-S.* Rev. ed. New York: Dodd, Mead & Company, 1967.

Simpson, R. W. *Airlife's Helicopters and Rotorcraft: A Directory of World Manufacturers and Their Aircraft.* Shrewsbury, England: Airlife Publishing, 1998.

Soviet/Russian Helicopters, http://www.royfc.com/links/acft_coll. Accessed November 2002.

Spencer, J. P. "Whirlybirds: A History of the U.S. Helicopter Pioneers." Seattle: University of Washington Press, 1998.

———. *Vertical Challenge: The Hiller Aircraft Story.* Seattle: University of Washington Press, 1992.

Swanborough, Gordon, and Peter M. Bowers. *United States Navy Aircraft since 1911.* London: Putnam, 1990.

Taylor, Michael J. H., ed. *Jane's Encyclopedia of Aviation.* New York: Portland House, 1989.

U.S. Air Force Historical Research Agency, http://www.au.af.mil/au/afhra/. Accessed December 2002.

Uttley, Matthew R. *Westland and the British Helicopter Industry, 1945–1960: Licensed Production vs. Indigenous Innovation.* London: Taylor and Francis, 2001.

Weinert, Richard P., Jr., and Susan Canedy, eds. *A History of Army Aviation, 1950–1962.* Ft. Monroe, VA: Office of the Command Historian, U.S. Army Training and Doctrine Command, 1991.

Young, Warren R., et al. *The Epic of Flight: The Helicopters.* Alexandria, VA: Time-Life Books, 1982.

CHAPTER 4

Vietnam, the Middle East, and the Face-off in Europe, 1961–1975

THE UNITED STATES IN VIETNAM

During the early 1960s, both NATO and Warsaw Pact countries brought into production rotorcraft designed, or initially flown, in the late 1950s. The UH-1 Iroquois came about as a result of Bell Helicopter's experimentation with turbine-powered helicopters and the U.S. Army's search for a larger and technically advanced medical evacuation helicopter. In the late 1950s and early 1960s, Bell engineers experimented with several modified Model 47s, including the Model 201, designated the XH-1 by the U.S. military. In 1954, Bell's engineers installed a 250-horsepower Turbomeca XT51 turbine engine in the experimental XH-1. Continental manufactured the turbine under license, and the craft became Bell's first turbine-powered helicopter. The Model 201 paved the way for the use of the Lycoming T53 turbine engine in the HU-1.

In order to select a new, turbine-powered helicopter, the Army proposed a competition for a new aircraft to replace the OH-13 in the medical evacuation role. The HU-1, with a longer, wider fuselage, accommodated two litters in a cross-wise configuration, but many at Bell believed the aircraft would ultimately excel as a troop transport and cargo carrier.

Interservice agreements of the time dictated that the Army not conduct the fly-offs of its aircraft. The Air Force or Navy conducted the evaluations according to specifications prepared by the Army. The Air Force selected the Kaman H-43 Husky, a turbine-powered helicopter with twin, intermeshing wooden rotors as competition for the HU-1. The results of the competition evaluations were almost identical, and the Air Force decided to let the secretary of the Army make the decision as to which rotorcraft to buy. The Air Force already utilized the H-43, and increased production would lower the unit cost of each helicopter. Also, Air Force officers believed they might influence the selection politically. Secretary of the Army Wilbur Brucker, after sifting through piles of technical data, asked his staff, "Bell—aren't they the ones who built that little bubble helicopter that saved so many of our boys in Korea?" When assured that was so, he replied, "I'm going with the company that did something for our boys in Korea," and on February 23, 1955, he awarded the contract to Bell Helicopter (Brown 1995, 98).

On October 20, 1956, the first of three Model 204 prototypes, designated XH-40 by the military, lifted off on its initial flight. It was powered by a Lycoming XT53-L-1 700-horsepower turboshaft, and Bell engineers fitted the helicopter with a conventional two-bladed main rotor and weighted stabilizer bar. The tailboom extended to a vertical fin that supported a two-bladed tailrotor. Horizontal stabilizers extended from both sides of the tailboom and were linked to the pilot's cyclic stick to increase controllability in forward flight. Behind the distinctive rounded nose and large windscreen sat the pilot and copilot. Aft of the cockpit, the cargo compartment widened to provide space for six passengers or up to four litters and a medical attendant. The 204 rested on skid gear designed for landings in rough terrain. Unfortunately, Larry Bell died the same day that the 204 lifted off on its maiden flight.

The preproduction YH-40 featured an upgraded T53-L-1A engine derated to 770 horsepower and a cabin lengthened by a foot. On June 30, 1959, the Army accepted the first of nine preproduction 204s, designated HU-1 Iroquois. The "HU," in military jargon, meant "Helicopter Utility," from which the HU-1 quickly received its legendary nickname, and the Iroquois would be forevermore known as the "Huey." As of 2004 more than 16,000 versions of the UH-1 were produced worldwide.

In 1959, Bell pilots flew a HU-1A to Fort Campbell, Kentucky, to demonstrate the new helicopter to officers of the 101st Airborne Division. Army officers told the Bell pilots to remove the litters and see how many combat troops the helicopter would carry. Lifting off with nine troops aboard, the helicopter duly impressed the paratroopers. Following the initial success of the Huey, the Army placed an order for 183 HU-1As, including 14 configured to train pilots to fly in Instrument Meteorological Conditions (IMC). The Army employed the HU-1s in Alaska, Europe, Panama, and Korea as medical evacuation helicopters. When the Army deployed the aircraft to South Vietnam in April 1962, many crews fitted the helicopters with improvised .30-caliber Browning or M-60 doorguns on flexible mounts. The Army also modified one HU-1A and employed it as a test aircraft for future versions of armed helicopters. Bell completed deliveries of the HU-1A in 1961.

In March 1961 the Army accepted the first improved HU-1B helicopter. A lengthened fuselage accommodated seven combat troops, or four stretchers, two sitting casualties, and a medical attendant. A T53-L-5 960-horsepower turboshaft and redesigned 44-foot rotor blades increased cargo capacity to 3,000 pounds. A T53-L-11 1,100-horsepower turboshaft replaced the L-5 in later HU-1Bs. The U.S. Navy borrowed a few of these Hueys from the Army and modified them for anti-submarine warfare (ASW) experimentation, including slinging depth charges under the fuselage. On April 4, 1963, Bell's commercial 204B received Federal Aviation Administration (FAA) certification, and foreign sales of both military and commercial models commenced. Bell produced at least 1,033 of all versions.

In Japan, Fuji Heavy Industries, which had been producing Bell Model 47s, including its own version the KH-4 since 1952, acquired a contract to produce ninety UH-1B helicopters for the Japanese Ground Self Defense Force (JGSDF). Fuji went on to manufacture fifty-five 204Bs for commercial sales. The Fuji model differed only in that the tailrotor was mounted on the right side instead of the left. In 1973, Fuji introduced the 204B-2 with a Kawasaki-Lycoming KT53-13B 1,400-horsepower engine and larger tailrotor that increased high-altitude performance. Subsequently Fuji also produced 145 UH-1H Hueys for the JGSDF with the Kawasaki-Lycoming KT53-13B 1,400-horsepower turboshaft.

General Hamilton H. Howze, recognized as the intellectual pioneer of U.S. Army aviation in the 1950s and 1960s, altered the employment of helicopters internationally. While director of Army aviation from 1955 to 1958, he initiated new principles of tactical doctrine and organizational structure of aviation units. In 1961, as chairman of the Tactical Mobility Requirements Board, Howze confirmed the necessity for developing airmobile theory and doctrine for Army aviation. The "Howze Board" revolutionized concepts of mobile warfare based on employing organic aviation assets. In 1963, as a result of the Howze Board recommendations, the Army formed the 11th Air Assault Division to test and validate airmobility concepts. The 11th Air Assault Division became the 1st Cavalry Division (Airmobile) prior to deployment overseas; in the next few years, it aptly demonstrated the requisite combined arms structure fundamental to airmobile operations on the modern battlefield.

During the early 1960s the United States began deploying helicopters to the Republic of Vietnam to support that country's efforts to combat the communist insurrection sponsored by the People's Republic of North Vietnam. On December 11, 1961, two companies of Piasecki H-21 Shawnees arrived by carrier in South Vietnam. Twelve days after their arrival the helicopters lifted more than 1,000 Army of the Republic of Vietnam (ARVN) paratroopers in the first helicopter combat assault in Vietnam. In April 1962, during Operation Shoofly, the first Marine CH-34Ds flew into South Vietnam. From late 1961 to early 1965, U.S. helicopter crews, expanding their knowledge by trial and error, instructed ARVN commanders in the tactical employment of helicopters. U.S. Army H-21, HU-1A, and Marine Sikorsky H-34 Choctaws, which Marine "grunts" lovingly called "Ugly Angels," evacuated casualties, supplied isolated garrisons, and provided rapid deployment of units to meet enemy threats. During its service in the RVN, in addition to military missions, U.S. Marine CH-34s also rescued more than 1,500 Vietnamese civilians from flood waters. By the end of 1964 the United States had over 250 helicopters in Vietnam. The success of these helicopter units forced North Vietnam to supply insurgent forces in South Vietnam with modern weapons and to escalate the movement of regular North Vietnamese Army (NVA) troops and supplies into the RVN by the Ho Chi Minh Trail network.

In 1963 the United States began supplying the Vietnamese Air Force (VNAF) with H-34s and upgraded utility and transport helicopters as newer types became available. VNAF pilots received training by U.S. advisors on the same tactics and procedures implemented by U.S. forces. U.S. requirements for helicopters precluded VNAF units from acquiring enough aircraft to institute independent airmobile operations to any large degree.

In September 1962 a triservice agreement by the U.S. military standardized the designations of all U.S. military helicopters. The

HU-1 became UH-1 meaning "utility helicopter." "CH" defined cargo helicopters, "OH" observation, and "AH" designated attack helicopters.

Initially, tactical heliborne transportation received little artillery or tactical air support, and U.S. advisors in South Vietnam soon realized that they needed additional firepower to conduct airmobile operations. In September 1962 fifteen Bell UH-1B Iroquois ("Hueys"), fitted with the XM-6E3 armament system, arrived at Tan Son Nhut, South Vietnam. The armament system included four M-60 machine guns mounted in pairs on outriggers outboard of the cargo doors, and a 7-round 2.75-inch (70-mm) Folding Fin Aerial Rocket (FFAR) rocket pod affixed to each side of the aircraft. The pilot aimed the machine guns with a sight connected to a hydraulic actuator that flexed the guns 70 degrees to the left and right. The U.S. Navy also flew a few UH-1B gunships, nicknamed "Sea Wolves," in support of riverine operations. The gunships scouted ahead of river patrol boats (PBRs) searching for ambushes and provided fire support for the PBRs in the event of an attack. Helicopters, nevertheless, remained vulnerable to enemy ground fire. During the battle at Ap Bac near Saigon in January 1963, Viet Cong gunners downed four H-21s and one armed Huey. Regardless of losses, however, by experimentation in actual combat, and applying lessons learned by the French in Algeria, U.S. pilots wrote the book on tactical employment of armed helicopters.

The UH-1B, nonetheless, lacked sufficient power to carry a large load of ordinance and stay abreast of the troop transport helicopters. In September 1965, to overcome those limitations, Bell introduced the UH-1C, designed for the gunship role. The "C" model featured an uprated T-53-L-11 engine, larger fuel tanks, and a new rotor system. The "540" rotor system eliminated Bell's usual stabilizer bar and replaced it with an electro-mechanical system called Stability Control Augmentation System, or SCAS. Bell engineers also lightened the rotor blades and increased their cord to 27 inches, providing more lift, speed, and maneuverability. The military designated a 1969 version of this Huey, modified with a more powerful T-53-L-13 turboshaft derated to 1,400 horsepower, the UH-1M. Bell manufactured about 750 "C" models, most later converted to the "M" version. Eighteen countries bought "B" and "C" models for their military services. The technology utilized in the "C" model became a database for the AH-1 Cobra, a helicopter designed specifically as a heavily armed attack helicopter.

The Army equipped at least three UH-1Ms with the Hughes Corporation's INFANT (Iroquois Night Fighter and Night Tracker) system, which made use of a low-light-level television (LLLTV) and infrared searchlight to aim the M21 armament subsystem. These "Mike Model" gunships carried the AN/AAQ-5 Forward Looking Infrared (FLIR) fire control system, a component developed for use on the AH-1G SMASH Cobra (Southeast Asia Multi-Sensor Armament Subsystem for Huey Cobra). The AN/AAQ-5 produced a televised thermal image, enabling the crew to detect, identify, and fire on ground targets during day or night operations. The 1st Cavalry Division (Airmobile) evaluated the aircraft from December 1969 to February 1970, but the Air Force AC-130 Specter gunship proved more effective for night operations.

The "B" and "C" Huey gunships came equipped with a variety of armament systems. The XM-3 consisted of two rectangular 24-tube rocket pods and no guns. The XM-16 used the new cylindrical XM-158 7-tube rocket pods along with the quad M-60 machine guns. The M-5 added a nose turret mounting an M-75 40-mm automatic grenade launcher with a rate of fire of 220 rounds per minute. The XM-21 replaced the XM-16, retaining the XM-158 rocket pods, but included two GE—134 six-barreled 7.62-mm miniguns, each with a rate of fire of 2,000 rounds per minute, instead of the M-60s. Army aviators called a Huey gunship equipped with the XM-3, carrying a total of forty-eight 2.75-inch FFARs, a "Hog," and a Huey gunship equipped with the M-5 as well, a "Heavy Hog."

In July 1960 the U.S. Army awarded Bell Helicopter, which became part of the Textron Corporation in mid-1960, a contract for seven prototypes of the YUH-1D, an extended version of the Huey and designated the Model 205 by Bell. The "D" model first flew on August 16, 1961, and went into limited active service in August 1963. A Lycoming T53-L-11 1,100-horsepower turboshaft drove a larger 48-foot rotor, increasing the cargo load to 4,000 pounds and passenger capacity to twelve to fourteen, or six litters and a flight medic. The UH-1D's larger self-sealing fuel system included provisions for external auxiliary fuel tanks. Two windows in the 205's cargo doors easily differentiated the aircraft from the single windows of the 204. A pintle-mounted M-60D machine gun in each cargo door became the standard self-defense armament of all Hueys. During the mid to late 1960s the UH-1D became the Army's primary troop transport and medevac helicopter, while the smaller 204 variants generally performed the gunship role. The Army bought a total of 2,008 UH-1Ds.

In February 1963, Bell Helicopter test pilots flew the first UH-

1E, ordered the previous year to provide the U.S. Marine Corps with an assault support helicopter. Like the "B" model, the UH-1E ordered for the Marines had an external rescue hoist, a rotor brake to quickly stop the free-wheeling rotor blades and hold them in place during shipboard stowage, and avionics particular to the Marine Corps mission. In February 1964 the USMC took delivery of the first of 250 UH-1Es. The next year all production "E" models left the Bell factory with the "540" rotor system of the Army "C" model. When the Marines armed some of their Hueys with .30-caliber machine guns and 2.75-inch rockets to escort the vulnerable CH-34s, Marine pilots began complaining about "shoddy" workmanship by Bell employees. When Bell technicians investigated the problems, they discovered that Marine pilots routinely dove in to attack targets at 165 knots, 25 knots above the maximum airspeed recommended for the helicopter, and then violently pulled up to escape ground fire. The Marine pilots simply overstressed the airframe by pulling too many "Gs." Knowing that the Marine pilots would follow their informal axiom, "Don't get caught low, slow, and stupid. If you do, you will be dead," Bell representatives just recommended that maintenance personnel securely tie down the engine cowlings and "keep on doing what they were doing" (Brown 1995, 111).

In June 1963 the U.S. Air Force contracted for a helicopter to perform missile site support duties, ordering the UH-1F. Bell engineers installed a General Electric (GE) T58-GE-3 1,290-horsepower turboshaft in the fuselage of the UH-1B to drive a UH-1D rotor system. On February 20, 1964, the UH-1F lifted off on its initial flight, and deliveries began in September. The USAF purchased a total of 146 UH-1Fs, with several modified as "UH-1P" psychological warfare aircraft, which carried loudspeakers over the jungles of Vietnam in an attempt to persuade Viet Cong and NVA soldiers to surrender. The USAF also bought about twenty "TH-1Fs" to train pilots in instrument flying and rescue hoist techniques.

In Vietnam, U.S. planners divided heliborne combat assault operations, soon called air assaults, into three phases: en route, approach, and landing. Armed helicopters proved most effective during the landing phase—after a few missions, ground fire hits on transport helicopters dropped from 0.011 hits per flying hour to 0.0074 for escorted aircraft. Hits on unescorted helicopters doubled during the same period. Suppressive fire delivered by armed helicopters proved very effective in reducing the amount and effectiveness of enemy fire on transport helicopters.

As a result of initial combat experience, a platoon of five to seven armed helicopters formed an escort for twenty to twenty-five troop-carrying helicopters. As transportation helicopters approached a landing zone (LZ), gunships began racetrack, or similar patterns, on each side of the landing helicopters. The gunship pilots fired rockets and machine guns on enemy concentrations while door-gunners directed suppressive machine gun fire as the armed helicopters broke away from enemy positions.

Military Assistance Command Vietnam (MACV) instituted Eagle Flights that included an armed Huey, piloted by the U.S. aviation commander and carrying the ARVN troop commander. This command and control (C&C) aircraft flew at an altitude above the reach of small arms, while the air mission commander directed seven to ten transport helicopters, dubbed "slicks" because they lacked external gun mounts, toward the LZ. A flight of five gunships escorted the formation to provide fire support for the insertion. A medevac helicopter trailed the formation to extract any casualties. Eagle Flights provided immediate response to targets of opportunity and could easily be melded into one large airmobile operation. Eagle Flights became the basis for airmobile concepts employed by U.S. combat units arriving in Vietnam in 1965.

In August and September 1965 elements of the 1st Cavalry Division (Airmobile), the first such division in the world, began to arrive at An Khe. The division's organization radically differed from that of a standard infantry division. The 1st Cavalry contained 434 helicopters, divided into two battalions of assault helicopters, a battalion of attack helicopters, a battalion of assault support helicopters, an Aerial Rocket Artillery (ARA) battalion (the first such battalion in the Army), and an air cavalry squadron. The division's capabilities included inserting one-third of its combat power at a time into terrain inaccessible to normal infantry vehicles. To support the large number of aircraft in the division, an Aviation Maintenance Battalion augmented normal Division Support Command. In June 1968 the 101st Airborne Division, deployed to RVN in 1965, received a change of organization and became the second airmobile division in the U.S. Army.

In September 1967 the first of a more powerful line of Hueys rolled out of Bell's Hurst, Texas, manufacturing facility. Continually seeking to improve the UH-1's performance in the hot, humid environment of Vietnam, Bell installed the more powerful T53-L-13 engine in the new helicopter. Unfortunately, the aircraft's transmission limited the engine to 1,400 horsepower. The new Huey measured 57 feet, 10 inches overall, with a 48-foot rotor diameter. With a maximum gross weight of 9,500 pounds, the aircraft reached a top speed of 120 knots, a service ceiling of 12,600 feet, and a maximum range of 280 nautical miles. It was designed to carry up to ten troops in Vietnam's density altitudes (the combination of humidity and pressure altitude), as well as two M-60 doorguns, armor for the crew, critical engine components, and a fuel system designed to prevent postcrash fires; that, however, reduced the aircraft loading (ACL) to six to eight combat loaded infantrymen. Conversely, a confirmed report stated that a single Huey lifted twenty-five troops out of a hot LZ. The Huey gained a reputation as a rugged, dependable helicopter, more than once returning to base with a hole in the rotor blades large enough for a pilot to stick his head through.

Increased power made the "H Model" a more effective medevac, or "Dustoff," helicopter. The universal sobriquet Dustoff for medevacs came from the U.S. Army 5th Medical Detachment call sign, selected by Medal of Honor winner Major Charles L. Kelly. Kelly earned great respect for his perseverance and cool demeanor under fire. When told to leave a hot LZ because of heavy enemy fire, Kelly calmly responded over the radio, "When I have your wounded."

The Army modified 220 UH-1H to UH-1V medevac aircraft, which included a rescue hoist and a bullet-shaped "jungle penetrator" device designed to lift casualties from the triple canopy jungle of Vietnam. Capable of carrying six litters, the medevac versions also included advanced avionics and navigation systems for night and inclement weather operations. Unarmed Dustoff helicopters experienced a loss rate of about three times that of other helicopters in Vietnam; the red crosses painted on the sides and nose of the Hueys probably made better targets for the NVA and VC gunners. The USAF also bought thirty similar HH-1Hs for short-range search and rescue (SAR).

Part of the weight problem with Hueys in Vietnam related to the manner in which Army pilots flew the aircraft. Although engine instruments indicated normal operations, they seemed to produce less and less power. Pilots routinely landed in flooded rice paddies, which allowed about 100 pounds of water and silt to seep in under the cabin floor. The water drained out, but after inspections some Hueys had as much as 300 pounds of dirt and silt trapped under the cargo deck.

The UH-1H remained in production for more than twenty years. The U.S. Army purchased a total of 3,573 UH-1Hs, with many remaining in service in 2004, mainly with Army Reserve/National Guard units. State-of-the-industry avionics, improved composite rotor blades, chaff-flare dispensers, and infrared countermeasures equipment maintained the Huey's viability. The Army also modified three UH-1Hs as EH-1H "Quick Fix" aircraft, designed to jam or intercept enemy radio transmissions. Excluding those manufactured under foreign license, Bell produced 1,372 UH-1Hs for export sales.

In 1965, Bell, utilizing company funds, experimented with a twin-engine Model 208 "Twin Delta" prototype, which was a UH-1D fitted with a Continental XT67-T-1 module. In this modification two T72-T-2 turboshafts drove a single transmission, turning Bell's standard semirigid, two-bladed main rotor system. In 1968, Bell, the Canadian government, and Pratt & Whitney Canada (PWC) opened negotiations that led to an agreement in 1969 to build a twin-engine version of the Model 205. In 1969, Bell modified a UH-1D with a new PWC "Twin Pac" containing two PWC PT6 turbines linked to a common reduction transmission. This prototype led to the production of the Model 212, or UH-1N "Twin Huey," essentially a UH-1H modified with the PWC T400-CP-400 Twin Pac, producing 1,530 horsepower. Each PT6T-3 was capable of 900 horsepower, but the combined 1,800 horsepower overwhelmed the 48-foot, 2-inch rotor system. In the event of an engine failure, however, the transmission allowed the remaining operational engine to deliver its full 900 horsepower to the rotor system. The second engine also increased the machine's maximum gross weight to 10,000 pounds and increased airspeed by about 10 knots. In 1970, Bell delivered the first of seventy-nine UH-1Ns to the USAF, which used the helicopters for special operations and SAR. By 1978 the U.S. Navy and Marine Corps accepted 221 UH-1Ns, which particularly suited both services' mission requirements because of the greater overwater flight safety provided by the twin-engine configuration. The Marine Corps converted two standard UH-1Ns to "VH-1N" VIP transports and then ordered six more production models of this version.

On May 3, 1971, the Canadian Armed Forces, which had funded much of the development of the 212/UH-1N, received the initial shipment of fifty aircraft ordered. Within a year Bell completed the contract. The CAF originally designated the helicopter the CUH-1N but later changed the designation to CH-135. The enhanced safety provided by the twin-engined 212 proved to be an excellent selling point for commercial operators working in offshore gas and oil exploration, resulting in sales to several commercial operators, in addition to foreign military services. Bell and its licensees produced more than 1,000 models of the 212.

In 1952 the Hughes Tool Company began experimenting with rotorcraft. Following the tradition of Howard Hughes's "Spruce Goose," the XH-17 "Flying Crane" was powered by two GE turbines that turned a huge two-bladed main rotor system measuring 133 feet, 10 inches in diameter. The aircraft stood 30 feet tall on four long legs designed to straddle large cargo. Company records indicated that the large machine could lift off at a gross weight of 40,900 pounds, but, like the Spruce Goose, the company built only one of the aircraft; it flew only once, in October 1952. The company continued experimentation with helicopters, and in 1956 it concluded Hughes's first contract to provide the U.S. military with a helicopter. The Army designated the Model 200/300, the TH-55 Osage, and bought a total of 792 of the small two-place machines as a primary helicopter trainer. Much smaller than the company's previous endeavors, this craft had a rotor diameter of 26 feet, 10 inches and a total length of 30 feet, 10 inches. A single Lycoming HIO-360 190-horsepower piston engine drove the transmission and tailrotor through a series of four belts. With a maximum gross weight of 2,050 pounds the machine reached a top speed of only 80 mph, but it laid the groundwork for the very successful Hughes 369.

Hughes continued to produce the Model 269/300 for several years, but in November 1983 it subcontracted all Model 300 production to Schweizer Aircraft Company in Elmira, New York, and the company purchased all rights to the 269/300 series in November 1986. Schweizer produced several specialized variants of the little helicopter, including agricultural versions, and the "Sky Knight," with a 420-horsepower Allison 250-C20 turboshaft for police duties, and the TH 300C dual-control version for foreign military services. Turkey, New Zealand, Norway, Liberia, Hungary, Japan, and Argentina all utilized variants of the Model 269/300. In 1990, Schweizer Aircraft also proposed a turbine-powered Model 300 to replace the Army's TH-55.

In the early 1960s, the U.S. Army began a search for a turbine-powered Light Observation Helicopter (LOH) to replace the piston-engined Bell OH-13s. Several companies provided bids for the prototype, but Hughes Aircraft won the contract with the Model 369, which became the OH-6 Cayuse. First flown on February 27, 1963, the agile little aircraft with the three-finned tail was soon being called the "Loach" by U.S. soldiers. The unique teardrop-shaped fuselage earned the OH-6 the sobriquet "Flying Egg," but the shape and internal bulkheads provided an exceptionally crashworthy aircraft, especially after Hughes installed self-sealing, crash-resistant fuel tanks to prevent postcrash fires.

The large Plexiglas windscreen offered the pilot excellent visibility, and the four-bladed fully articulated rotor system provided much more maneuverability than the semirigid rotor of the OH-13. The Allison T63-A-5A 250 turbine produced 317 horsepower, giving the OH-6 more speed, 130 knots, and allowed the aircraft to carry up to five passengers, or 1,000 pounds of internal cargo. In 1966 the OH-6 set the first of its twenty-three international records by covering 1,923 nautical miles in a single straight-line flight. In 1967 the Loach first appeared in Vietnam performing such duties as command and control, liaison, fire direction, light utility, and reconnaissance. In the air cavalry role, the crew usually consisted of a pilot and observer; the helicopter was fitted with an M-27 7.62-mm minigun system, and the observer was armed with an M-60 machine gun. The Army eventually ordered 1,420 OH-6s, which were delivered between 1966 and 1970.

The OH-6 proved so effective that the Japanese Self-Defense Force let a contract with Kawasaki Heavy Industries to produce a version of the helicopter. KHI built the export version of the OH-6, the Hughes 500M Defender, designated the OH-6J in Japan. KHI produced a later version with a larger five-bladed rotor system designated the OH-6D. The Japanese military purchased 120 OH-6Js and 62 OH-6Ds. Kawasaki also marketed the 369HS and 369D for EMS, police, and agricultural services.

In 1964, to provide battlefield commanders with a fast, heavily armed, technologically advanced, and less vulnerable attack and reconnaissance helicopter, as well as to support airmobile tactics and air cavalry operations in Vietnam, the U.S. Army initiated the Advanced Aerial Fire Support System (AAFSS) Program. Lockheed Aircraft won the competition with the AH-56A Cheyenne, but an immediate conflict erupted between Army procurement and operational officers. Army planners believed the AAFSS, which lay in the future, to be the ultimate solution to all attack helicopter concerns. Operational officers, however, demanded an advanced gunship to deploy to Vietnam immediately. When Colonel George P. Seneff, chief of Army aviation, told General Harold K. Johnson, Army chief of staff, that our "troops are dying in Vietnam now, not in the future," Johnson decided to purchase an intermediate attack helicopter.

Bell Helicopter's recent developments placed the company in an ideal position to fill the Army contract for an intermediate attack helicopter. In 1963, Bell produced a prototype Model 207 Sioux attack helicopter, which combined a Model 47J tail assembly with a tandem-seat Plexiglas cockpit. The helicopter incorporated short shoulder-mounted wings and a chin-mounted turret with two machine guns. A single Lycoming 260-horsepower piston engine turned the typical Bell semirigid rotor system. For the new attack design, Bell engineers incorporated the lessons learned with the Model 207, the airframe of the OH-4 Sioux Scout, which had lost the LOH competition to the Hughes OH-6, and the Lycoming T53-L-13 1,400-horsepower engine to produce a new prototype attack helicopter. To the airframe and engine, the engineers added the transmission, "540" rotor system, and weapons technology of the UH-1C gunship to create the Bell Model 209, a helicopter engineered specifically as an attack helicopter. On its maiden flight in September 1965, the 209 demonstrated an airspeed of 170 knots, almost twice that of the UH-1B and Cs.

In March 1966 the Department of Defense (DoD) contracted to buy 1,100 of the new helicopters, designated the AH-1 Cobra by the U.S. military. The Cobra differed radically in appearance from previous Bell designs. The pilot and copilot/gunner sat in an armored step-seating tandem cockpit, the gunner in front and the pilot in the rear. The thin 38-inch-wide fuselage and jet fighter–type canopy offered the crew excellent visibility and reduced the aircraft's frontal silhouette to enemy gunners. Stub wings and a nose-mounted chin turret provided hard points for a variety of weapons systems. Fixed skids replaced the retractable landing gear of the first Model 209.

The AH-1G Cobra, or "Snake" as it became known to Army aviators, carried 2.75-inch FFARs in either M158 7-tube or M200 19-tube rocket pods affixed to the wings. The chin-turret mounted the M134 7.62-mm minigun and M129 40-mm grenade launcher. Several air cavalry units armed Cobras with miniguns in M18/M18A1 gun pods installed on both wings, or the M35 armament subsystem, which included an M195 20-mm automatic cannon fixed to the left wing. On selected missions the Ah-1 carried the XM118 smoke grenade dispenser. All AH-1s mounted the M73 reflex sight for the pilot to fire the weapons systems. The Army equipped a limited number of Cobras with the CONFICS (Cobra Night Fire Control System) and the SMASH systems to provide the gunships with a night firing capability. A few early-model AH-1G/AH-1Q Cobras mounted either two M134 miniguns or two M129 grenade launchers in the turret, but problems with the ammunition feed systems caused the Army to abandon the twin-gun configurations.

In late 1966 and early 1967 the AH-1G began to replace most of the gunships in air cavalry and aerial rocket artillery (ARA) units, providing more flexibility and firepower. Air cavalry troops (company) organized their helicopters into three platoons: the attack helicopters in the "Red" Aerial Weapon Platoon, the scouts in the "White" Aerial Scout Platoon, and the Hueys in the "Blue" Aerial Rifle Platoon. A combination of a scout and attack helicopters became "Pink" teams, and the Cobras, operating with scout helicopters, especially the OH-6, became one of the most effective combat multipliers in the U.S. Army. The scouts flew with their skids brushing the trees, and the attack helicopters dove in and blasted any enemy formations discovered. Many commanders in RVN estimated that their air cavalry units initiated as many as 90 percent of their engagements with the VC and NVA.

ARA units, in contrast to other gunships, operated under direction of artillery officers. Divisional or separate artillery commanders employed these helicopters to supplement conventional tube artillery. When called, ARA platoons communicated on fire direction radio frequencies, and the helicopter's rockets were adjusted much like conventional artillery. ARA's great advantage lay in its long-range and heavy firepower. ARA provided direct fire support to units out of range of conventional artillery. The Bell AH-1G Cobra carried up to seventy-six rockets, and with 17-pound warheads it delivered the same initial firepower as a battalion of 105-mm howitzers—devastating firepower for one helicopter.

The U.S. Marines also operated helicopter gunships in Vietnam and incorporated attack helicopters into their future tactical planning for both amphibious and vertical ship-to-shore operations. The Marines envisioned attack helicopters providing armed escort for troop-carrying helicopters; close-in-air support for beachheads and landing zones, including an antiarmor role; armed reconnaissance; and self-protection against armed enemy helicopters. In May 1968 the Marines ordered their own version of the Cobra. A preponderance of overwater flights by the USMC demanded a twin-engine AH-1, therefore Bell Helicopter, by this time a division of Textron, developed the AH-1J Sea Cobra. Constructed around the AH-1G airframe, the Sea Cobra included a twinpac of Pratt & Whitney T400 900-horsepower turboshafts, thus increasing available power. Bell also increased the AH-1J's firepower by installing a three-barrel XM-197 20-mm cannon in a new nose turret. The Sea Cobra included improved dynamic components, avionics, and a rotor brake for shipboard operations.

While the USMC awaited the development and production of the first forty-nine AH-1Js, the Corps bought thirty-eight AH-1Gs from the Army. Army instructors conducted the initial training of Marine pilots, and, in April 1969, VMO-2 became the first operational Marine Cobra unit in Vietnam. In December the USMC transferred the AH-1Gs to HML-367. The same month flight evaluations began on the Sea Cobra, and the first AH-1Js entered service in February 1971, commencing combat operations the following month. Marine AH-1Js, including those of HMA-369, flew combat missions in Southeast Asia (SEA) until the final withdrawal of the United States in 1975. The USMC bought a total of sixty-seven AH-1Js, which remained the Marines' primary attack helicopter for several years. The Corps transferred the AH-1Gs to reserve helicopter attack squadrons.

As early as 1966 the U.S. Army adapted a few UH-1Bs to carry up to six French-designed SS-11B wire-guided antitank missiles to engage hard targets. Adopted by the Army as the AGM-22B and mounted on the M-22 missile subsystem, the missiles were never a popular weapon. Gunners guided the SS-11 by eyesight, tracking the missile by a flare in its tail and adjusting the missile's flight with a joystick. Wires, spooled out behind the missile, transmitted course corrections from the operator to the missile in flight. The system required highly trained gunners and a "fairly benign combat environment," and, as both were in short supply, the accuracy of the AGM-22 proved illusory at best (Williams 2003).

In 1966 the Army Aviation Test Board also armed two UH-1Bs with the new Hughes BGM-71 Tube-launched, Optically tracked, Wire-guided" (TOW) antitank missiles as part of the AH-56 Cheyenne test program, but the Hueys were placed in storage when funds were reduced for the AH-56. When the NVA began to introduce Soviet-manufactured tanks into RVN, the Army reactivated the TOW-equipped Hueys and conducted extensive firing tests before deploying the helicopters to Vietnam. Much more advanced than the AGM-22, the TOW provided far greater accuracy. The gunner simply maintained the sight on the target while the fire-control system guided the missile to the target. During the NVA "spring offensive" of 1972, Army pilots fired eighty-one TOW missiles and recorded fifty hits. In contrast, of the twenty-one SS-11 missiles fired during the same period, only three hit their targets.

AH-1 Cobras also destroyed several tanks during the siege of An Loc in 1972. Responding to a U.S. Army advisor's request to support the ARVN troops against an overwhelming NVA attack, the Cobras dove through intense ground fire and destroyed three T-54s with 2.75-inch high explosive anti-tank (HEAT) rockets, halting the NVA attack. During the next few days, Army Cobras destroyed twenty

T-54s with their rockets. Unfortunately, eight of the thirty-two Army aviators involved in the action died in combat. At an after-action review in the Pentagon, Air Force operational specialists claimed that the 2.75-inch rockets were not accurate enough to hit tanks. The Air Force contended that attack helicopters would never be a viable weapon on mid- to high-intensity battlefields. When asked how close the Cobra pilots fired their rockets, Lieutenant Colonel Bob Molonelli said, "About one hundred yards. You told us to kill them and we did." The Air Force officers failed to account for the tenacity of Army aviators in their calculations. The success of TOW and rocket-firing helicopters in Vietnam substantiated the practicality of the antiarmor helicopter and inaugurated an intensive research program by the U.S. Army on how to best employ the versatile weapons systems (Williams 2003).

Lockheed Aircraft designed the AH-56A Cheyenne to meet U.S. Army requirements for the AAFSS, which began in 1964. On May 3, 1967, Lockheed rolled out the first of ten prototypes of the rigid-rotor Cheyenne. A single General Electric T64-GE-16 3,435-horsepower turbine powered a four-bladed rigid main rotor and anti-torque tailrotor, as well as a three-bladed pusher propeller mounted at the rear of the unusual aircraft. At 100 knots the aircraft's stub wings assumed approximately 80 percent of the lift, allowing the Cheyenne to reach a top speed of 214 knots in level flight. Lockheed engineers provided armored, tandem seating for a pilot and copilot/gunner in a jet fighter–like cockpit and armed the Cheyenne with an innovative XM-112 swiveling gunner's station, linked to a rotating belly turret containing an XM-52 30-mm automatic cannon and a chin turret with either an XM-51 40-mm grenade launcher or an XM-53 7.62-mm Gatling gun. An advanced-fire control computer, which included a laser range finder, controlled the guns as well as the eight TOWs and thirty-eight XM200 2.75-inch rockets carried in launchers mounted under the wings. Following the 1966 decision to buy the AH-1G, DoD decreased funding to the Cheyenne program. On August 9, 1972, because of delayed development, technological difficulties, rising costs, mishaps, and the appearance of two competitive helicopters developed by Bell and Sikorsky, DoD officially canceled the AH-56.

In 1970, with problems increasingly plaguing the AH-56 program, Sikorsky independently developed the S-67 Blackhawk attack helicopter prototype. Sikorsky's intermediate AAFSS helicopter was an armed, modified version of the company's proven S-61, H-3 Sea King. Two General Electric T58-GE-5 1,500-horsepower engines turned a five-bladed main rotor. New night vision systems provided targeting information to a Tactical Armament Turret (TAT-140) containing a 30-mm cannon; additional armament included up to eighteen 130-mm TOW missiles, and either 114 2.75-inch rockets or AIM-9 Sidewinder air-to-air missiles. Sikorsky engineers installed speed brakes on the trailing edges of the aircraft's short wings to improve maneuverability and a retractable landing gear to increase speed. Although the aircraft set a number of new records during a rigorous test program from 1970 to 1974, the Army judged the Blackhawk unsatisfactory for its purposes. On December 14, 1970, the S-67 established a world speed record by flying at 249.53 knots over a 1.86-mile course. In 1974, near the end of the test program, Sikorsky replaced the conventional tailrotor with a ducted fan, allowing the S-67 to reach a speed of 264.7 knots in a dive. Unfortunately, in a demonstration roll during an air show at Farnborough, England, the S-67 struck the ground, destroying the aircraft and killing both pilots.

In 1971, Bell Helicopter introduced the model 309 King Cobra prototype, also company financed, to contend for the AAFSS contract. One of only two prototypes crashed, but the other flew in comparative trials with the Lockheed AH-56A Cheyenne and the Sikorsky S-67 Blackhawk. Army officials, however, determined that none of the aircraft met their requirements. A single Lycoming T55-L-7C 2,850-horsepower turboshaft powered the King Cobra. Bell designers equipped the King Cobra with a laser day and night sight, infrared fire control system, night vision TV, and armament similar to that of the Cheyenne and Blackhawk.

Kaman also entered the fray with an armed version of the Seasprite. The company modified six aircraft with a four-bladed tailrotor and tweaked the engines to raise the maximum gross weight to 12,500 pounds to offset the added armor plating. Kaman altered the nose section with a turret-mounted 7.62-mm minigun and installed two other miniguns to the sides of the aircraft. The modified Seasprite created no real interest, however, and Kaman dropped the project.

In the early 1970s the Army initiated the Improved Cobra Armament Program (ICAP), which resulted in the AH-1Q antiarmor version of the Cobra. The AH-1Q incorporated the XM65 TOW/Cobra missile subsystem. The XM-65 included a telescopic sight unit (TSU) mounted in the aircraft's nose that magnified targets by thirteen times and allowed gunners to pinpoint their targets. The AH-1Q carried up to eight Hughes BGM-71 130-mm TOW antitank missiles in paired, stacked launchers mounted on the outboard wing pylons. Depending on the mission, commanders had the option of installing either the M158 7-tubed or M200 19-tubed 2.75-inch FFAR rocket pods on the inboard pylons. In 1973 the Army deployed a modest number of AH-1Q Cobras to Vietnam, but the heat and humidity, resulting in high-density altitudes, prevented the aircraft from carrying full ammunition and fuel loads, initiating a search for a more powerful attack helicopter.

Bell Helicopter, Textron, produced another helicopter that saw extensive service in the RVN. In 1967 the U.S. Army, because of escalating prices for the OH-6 Cayuse and its spare parts, reopened bids for the LOH program. The Bell civilian Model 206A Jet Ranger, derived from the unsuccessful OH-4, won the competition the next year, which resulted in an order for 2,200 aircraft. Designated the OH-58A Kiowa by the Army, the helicopter had the usual Bell all-metal, two-bladed, semirigid main rotor system, but without the stabilizer bar, and a two-bladed "Delta-hinge" tailrotor. Powered by a single Allison 250-C10 T63-700 317-horsepower gas turbine engine, the Kiowa reached a maximum airspeed of 120 knots and carried a M-134 7.62-mm minigun in the air scout role. Designed as a single-pilot aircraft, the OH-58A could carry three passengers or 400 pounds of cargo to a range of 300 nautical miles and a service ceiling of 18,900 feet. The Navy evaluated the Jet Ranger as an ASW platform, ordering prototypes with folding blades and armed with an acoustic torpedo. The Navy rejected the ASW version but bought nearly 200 TH-57s for pilot training. Because of the high demand for helicopters during the Vietnam years, Bell signed a multiyear contract with Beech Aircraft to produce OH-58 fuselages and established production facilities in Canada, which manufactured most of the later OH-58/Jet Ranger models.

The helicopter received FAA commercial certification in October 1966, with civilian deliveries beginning in January 1967. Companies involved in offshore oil exploration demanded a helicopter capable of carrying more than the five passengers of the original 206A. In 1973, Bell engineers elongated the cabin of the 206 to accommodate seven passengers and designated the modification the Model 206L Long Ranger. Bell installed a spray rig for dusting crops in an agricultural version of the machine. Companies worldwide made use of the faster and more powerful 206L in sundry commercial applications, and thirty-five countries employed the military versions of the helicopter. In addition to the United States and Canada, the Jet Ranger was also manufactured in Australia and Italy.

Throughout the 1960s, Stanley Hiller continued to produce the OH-23 Raven and experimented with new designs. The U.S. Army used the OH-23 in Vietnam as a light utility scout and as a training aircraft, as did the British Royal Navy and Royal Canadian Air Force. Several commercial operators utilized the helicopters as crop sprayers.

In 1960, Hiller Helicopter merged with Eltra Corporation (Electric Autolite Company) and introduced an improved Model 360, also known as the UH-12E and OH-23E Raven. The new model included an enlarged cockpit layout that placed a single pilot seat in front of a standard three-place passenger bench seat. The company offered retrofits of this modification to existing Hiller aircraft, and several operators took advantage of the opportunity to have their helicopters modified. Hiller also introduced a prototype Model 1099, a square-fuselage utility helicopter offering seating for six or an unobstructed cargo area accessed through a rear-opening door.

In 1961, Hiller submitted the Model 1100 as a contender in the LOH competition. The company delivered five aircraft to be evaluated as the OH-5A. Although Hughes won the contract, Hiller regarded the Model 1100 as too important to abandon and placed it into production for civilian and international military customers.

On May 5, 1964, Fairchild Aircraft acquired Hiller Aircraft and in September changed the corporate name to Fairchild Hiller. The new corporation phased out production of the UH-12 in favor of the new FH-1100. Between 1966 and 1974, Fairchild Hiller produced almost 250 of the machines.

In 1973, Hiller Aviation of Porterville, California, bought the rights to produce the UH-12 and resumed production of the UH-1E. Soloy Corporation devised a popular secondary market retrofit for the UH-12, the first of which flew on August 9, 1973. The modification included installing a 400-horsepower Allison 250-C20B turboshaft engine in the UH-12 airframe. Soloy converted 180 UH-12s, as well as some 220 Bell 47s.

Hiller Aviation also acquired the rights to the FH-1100 and prepared to manufacture that helicopter as well, but changes in ownership of the company delayed any significant production. In April 1984, Rogerson Aircraft Corporation acquired the company and, under the name Hiller Helicopters, then Rogerson Hiller, produced five RH-1100B Pegasus between 1983 and 1986. In 1990 the company resumed production of the UH-12E.

In 1959, Rudolph J. Enstrom established the RJ Enstrom Corporation at Menominee County Airport, Michigan, and developed a

light, three-seat civilian helicopter. In 1960 a prototype F-28 performed well enough for the company to seek certification of the little helicopter. Enstrom designed the F-28 with a fiberglass-enclosed
fuselage and engine compartment that housed a 180-horsepower
engine that powered a three-bladed main rotor. Despite some set-
backs the F-28 received its FAA certification on April 15, 1965, and
customer deliveries began the next January. Enstrom modified the
drive ratio of the F-28A to provide more power to the rotor, resulting
in several sales to Brazil and the Philippines. Developmental costs
precluded Enstrom from fitting a turbine engine to his little helicopter, but he continued to modify and upgrade the F-28A for several
years.

In 1989 the U.S. Army announced the Single Contractor Aviation
Training (SCAT) program, intended to contract a single provider for
Army flight training. Enstrom joined with Link Flight Simulation to
construct a turbine-powered helicopter to compete for the fleet of
trainers to be supplied by the contract winner. In December 1988,
Enstrom had installed an Allison 250B turbine in a 280FX airframe.
The resultant helicopter, designated Model 480/TH-28, embodied a
reshaped, wider forward fuselage that provided greater headroom
and an additional two seats. Bell Helicopter's TH-58 won the SCAT
project, and the TH-28 went without a buyer.

Enstrom, however, placed the civilian Model 480 into production,
where it gained a steady flow of orders. The company offered the
480 in a variety of seating arrangements for up to four passengers
and a pilot. An unusual staggered seat arrangement provided maximum legroom in the small cabin. Over the years Enstrom changed
ownership several times, but the latest partnership continued to
market the F-28F, 280FX, and the turbine-powered Model 480.

In 1962, MACV established a Search and Rescue headquarters
near Saigon, manned by personnel of the USAF Air Rescue Service,
to coordinate SAR operations in SEA. Not until 1964, however, did
the first of fourteen USAF detachments flying Kaman HH-43B
Husky aircraft arrive in Thailand. The Husky unit immediately began SAR operations, with orders to remain clear of the front lines.
With the call sign of "Pedro," the HH-43 proved very effective on
short-radius missions, but the Kaman machines lacked the range for
long-distance rescues. In many instances USAF crews resorted to
carrying fuel drums aboard on outbound legs and hand pumping the
fuel into the helicopter's tanks before completing their mission. Although the HH-43B set an altitude record of 32,840 feet in October
1961, and the operator's manual listed a service ceiling of 25,700

feet, the Husky crews experienced great difficulty operating in the mountains of Laos. Weight restrictions prevented the crews from carrying anything but a single M-14 or M-16 rifle for self-protection. Even when upgraded with the more powerful Lycoming T53-L1A 825-horsepower turbine, the HH-43 couldn't cope with mountainous terrain. Obviously, SAR missions demanded a larger, more powerful helicopter with increased range.

In 1956 the Piasecki Helicopter Corporation became the reorganized Vertol Aircraft Company, and a design team headed by Tom Peppler began work on a new company-funded, twin-engined, tandem rotor "flying banana." Typical of Piasecki's earlier designs, the three-bladed main rotors and transmission were mounted atop pylons at the front and rear of the aircraft. The two pilots sat side by side in a cockpit forward of the cargo compartment. The team envisioned the craft as carrying an entire infantry platoon and associated equipment and thus installed the engines on either side of the base of the rear pylon, leaving an uncluttered internal fuselage, with a hydraulic loading ramp at the rear. Aft sponsons contained the main wheels of a fixed tricycle landing gear.

On April 22, 1958, the Model 107 prototype, powered by 877-horsepower Lycoming T53 turbines, took flight for the first time. Three months later the U.S. Army ordered ten evaluation models, with the more powerful General Electric YT58 1,065-horsepower turbine engine and a larger rotor system, designating the aircraft YHC-1A. On August 27, 1959, the YHC-1A made its initial flight. Prior to the delivery of the first YHC-1A, however, the Army had ordered five Vertol YHC-1Bs (Model 114), a larger, more powerful aircraft that better suited the Army's requirements for a tactical medium-lift transport helicopter. Consequently the Army reduced the order for the Model 107 to only three machines. The Army utilized the aircraft for a short time to familiarize aviators with turbine-powered helicopters, then returned the aircraft to Vertol; one became the prototype for the Model 107-11 civilian version of the helicopter.

In March 1960, Boeing Aircraft Company acquired Vertol and introduced the Boeing/Vertol Model 107M. The aircraft, a modified YHC-1A, won a U.S. Navy competition for a new medium-lift transport helicopter, which resulted in an initial order for fifty of the HRB-1 helicopters. Purchased for the Marines, the first redesignated CH-46 Sea Knight commenced flight testing in October 1962. The main rotor blades folded, and provisions for emergency flotation allowed the Sea Knight to land and take off in light seas and remain afloat for up to two hours. Integral equipment included a rescue hoist, provisions for internal cargo handling, and a hoist capable of hauling 10,000 pounds of external cargo. The CH-46 entered fleet service in November 1964 and was employed as a troop transport/resupply helicopter in Vietnam, where Marine aircrews fitted 7.62-mm machine guns to fire out the cabin doors.

Between 1961 and February 1971, when production of the CH-46 ended, the Marines ordered more than 600 Sea Knights, with an initial purchase of 160 CH-46As, followed by an additional 266 CH-46Ds with uprated T-58-GE-10 engines and improved rotor blades. The Corps later bought 174 CH-46Fs, basically a "D" model with advanced avionics and electronics. Eventually the Marine Corps converted many of the earlier models to the CH-46E, installing more robust transmissions and rotor heads, composite rotor blades, and (Night Vision Goggles) NVG-compatible cockpits, providing the Marines with an all-weather, day-night assault helicopter. The Navy purchased twenty-four CH-46As and ten CH-46Ds. The USAF ordered a few "B" models for evaluation and the Navy the RH-46E minehunter for the same purpose. The Canadian Air Force ordered eighteen of the machines, designated CH-113, while Sweden bought fourteen aircraft designated HPK-4. Beginning in 1962, New York Airways bought a total of seven of the civilian version of the Sea Knight. As of 2000 the U.S. Navy and Marine Corps operated 291 Sea Knights. In Japan, Kawasaki Heavy Industries also built both civilian and military versions of the Model 107 under license: the KV-107/11-2 commercial passenger version, operated by Kawasaki, the Thai government, and New York Airways; the KV-107/11-3 minehunters; the KV-107/11-4 tactical transport for Japanese Ground Self-Defense Force; the KV-107/11-5 rescue version for the Japanese Air Self-Defense Force and the Swedish Navy; one KV-107/11-7 VIP transport for the Thai government; and the KV-107/IIA, modified specifically for hot climates and high altitudes.

On September 21, 1961, the Boeing-Vertol Model 114 lifted off on its maiden flight. The company produced the helicopter as a result of a September 1958 U.S. Army requirement for a medium transport helicopter capable of lifting a 4,500-pound load in all weather. In March 1959 the Army adjudged the modified Boeing-Vertol Model 107 winner of the competition. The Army placed an initial order for five units and classified the helicopter as the CH-47 Chinook. The machine's fuselage served as a large cargo compartment with the flight deck at the nose of the helicopter and a hydraulically operated ramp located at the rear. The cargo area accommodated forty-four combat-equipped infantrymen, or twenty-four litters plus two medical attendants, or large pieces of equipment or vehicles weighing up to 12,000 pounds. A hole in the forward floor of the cabin allowed items to be hoisted into or lowered from the helicopter. An external cargo hook at the center of the helicopter allowed the Chinook to carry large slingloads, up to 28,000 pounds on current models. Like its predecessors, the CH-47's design placed the three-bladed, counter-rotating, tandem rotor hubs atop pylons at the fore and aft ends of the aircraft. Each articulated rotor system, with blades composed of steel spars, aluminum honeycomb filler, and a plastic-reinforced fiberglass skin, measured 60 feet in diameter. The 67-horsepower auxiliary power unit (APU), combining transmissions for the turbine engines and all associated driveshafts, was installed on top of the fuselage as well. Fairings along the lower exterior of the helicopter housed large fuel tanks, the battery, and other electronic equipment. The landing gear consisted of four fixed struts, the aft set of paired wheels being steerable on the ground but locked in flight.

The Army settled on the Chinook as its primary medium assault transport helicopter, accepting the first delivery in August 1962, and equipped its airmobile test division, the 11th Airborne, with the CH-47A. When the airmobile division changed to the 1st Cavalry Division (Airmobile), it received the CH-47B with more powerful T55-L-7C engines. On October 14, 1967, the CH-47C appeared with even more powerful T55-L-11A 3,802-horsepower engines, stronger transmissions, a larger capacity fuel system, and an additional attaching point for external cargo. The next spring the U.S. Army began taking deliveries of the "Super C," as many soldiers called the upgraded Chinook.

In late 1965, in cooperation with the Boeing Corporation, the U.S. Army commenced testing an armed version of the CH-47 Chinook. Engineers mounted an array of weapons on the aircraft, including an M-5 40-mm automatic grenade launcher in a nose turret controlled and fired by the copilot. Pylons on either side of the helicopter accepted forward-firing weapons including a 20-mm cannon and 19-round 2.75-inch rocket pods. A modified fuselage allowed two doorgunners on each side of the cargo compartment to fire a 7.62-mm or .50-caliber machine gun situated on flexible mounts. An additional gunner position located at the rear loading ramp provided protection after the aircraft had completed its "gun run." One version of the ACH-47 even carried an experimental 105-mm gun. In addition to more than a ton of expendable ammunition, a ton of steel plating and heavily armored seats protected the aircraft crew and vital aircraft components from ground fire. The Army deployed a company of the armed Chinooks to the RVN, with the call sign of "Guns A-Go-Go," but the USAF AC-130 proved much more effective; the Army withdrew its armed Chinooks from combat.

In May 1970 the Boeing Corporation flew a company funded experiment with a modified Model 114. For the Model 347 Experimental Advanced-Technology Helicopter prototype, Boeing engineers stretched a CH-47A fuselage with a taller aft pylon, installed four-bladed main rotors, a retractable landing gear, and mounted an unconventional lift wing above the center of the fuselage. The experiment proved unfruitful, with no forthcoming orders for the unusual aircraft.

U.S. SAR tactics in Vietnam evolved from an operation developed by the German Luftwaffe during the Battle of Britain during World War II; they were refined during the Korean War. Four A-1Es called "Sandy" from their call signs, or A-7 jet fighters, broken into a high and low section of two aircraft each, escorted two HH-3 or, after 1967, HH-53 rescue helicopters into the Pickup Zone (PZ). The helicopters and two escorts orbited out of range of antiaircraft fire. The high section of fighters reconnoitered the site to determine enemy resistance and attacked antiaircraft weapons in the vicinity. The leader then called in the first helicopter with its two escorts. The second helicopter orbited out of danger, ready to replace the first if it was shot down or damaged. Many times the alternate helicopter swooped in to rescue both the downed helicopter crew and the object of the initial rescue mission. On several occasions, several U.S. aircraft and men were lost attempting to rescue downed airmen.

The helicopter most identified with U.S. long-range SAR operations in SEA first entered service as an ASW helicopter with the U.S. Navy. On March 11, 1959, Sikorsky Aircraft Corporation test pilots lifted off in the S-61, or H-3 Pelican, later known as the Sea King. The twin-engined, all weather Pelican became operational with the U.S. Navy in June 1961. The H-3, although designed primarily to detect, classify, track, and destroy enemy submarines, rapidly became a multipurpose ship or land-based aircraft with Navy, Marine, and USAF units. Powered by two GE T58-GE-10 1,250-horsepower turboshafts, soon upgraded to 1,400 horsepower, the H-3 cruised at an airspeed of 120 knots to a maximum range of 542 nautical miles and reached a service ceiling of 14,700 feet. The H-3 had a maximum gross takeoff weight of 21,000 pounds, with the capability of a 6,000-pound slingload on some models.

Of conventional design, the Sea King's five-bladed main rotor system measured 62 feet and its fuselage 54 feet, 9 inches. Designed for overwater flights the H-3 possessed emergency amphibious capabilities. The H-3 carried a crew of four and up to twenty-eight passengers in its transport configuration. For ASW operations the helicopter crew usually consisted of two pilots, two to three sensor operators, and up to three passengers. Dubbed the "Jolly Green Giant" by rescued aircrewmen, the USAF SAR version of the HH-3 usually carried two pilots, a door gunner, a parajumper (PJ), and a flight engineer who doubled as the rescue hoist operator. In 1967 two USAF Air Rescue and Recovery Service (ARRS) HH-3s demonstrated the viability of air refueling of helicopters by self-deploying from the United States to France in a flight of almost thirty-one hours (Pember 1998, 49).

Sikorsky produced two basic models of the S-61, the original with a watertight hull, which Navy personnel called "Big Mother," and later, the elongated version with aft cargo doors and a loading ramp for larger cargo. The United States, Great Britain, Canada, Japan, Italy, Spain, Egypt, and Malaysia used several variants of the Sea King: search and rescue HH-3 and SH-3D also used to pluck U.S. *Apollo* astronauts and their space vehicles from the ocean after splashdown, ASW SH-3, tactical transport CH-3, mine sweeping RH-3, VIP transport VH-3 (used by the U.S. Marines to transport the president), special operations MH-3, and commercial transport helicopter. Armament included a dipping sonar, sonobuoys, magnetic anomaly detectors, two MK 46 torpedoes for ASW, and at least two 7.62-mm door-mounted miniguns on SAR helicopters for self-protection. The aircraft contained cockpit instrumentation for all-weather operations, including both search and weather radars. Later models included Doppler radar navigation systems. In the early 1990s many Sea Kings received upgraded T-58-GE-402 1,500-horsepower engines and global positioning systems (GPS). The USCG HH-3F had the capability to fly 300 miles, hover for twenty minutes, and return to base with fuel in reserve.

Sikorsky built more than 1,100 S-61s, while Westland, Mitsubishi, and Agusta manufactured over 400 versions of the H-3 under license. Westland installed a pair of Rolls Royce Gnome H.1400 turboshafts and a Louis Newmark Mk 31 automatic flight control system in the British Sea Kings. The RN initially ordered fifty-six HAS1 Sea Kings, with 700(S) Squadron receiving the first for test and evaluation in August 1969. Testing resulted in the more powerful HAS2, and the HAS5 with a longer cabin to accommodate the

Sea Searcher radar. Westland provided the Egyptian Air Force with a twenty-one-seat Sea King utility transport, minus the external floats, called the "Commando." The British Royal Marines also ordered this version as the HC4, which conducted extensive combat operations in the Falklands War.

Westland also produced a completely self-contained SAR helicopter that carried a crew of four, nine stretchers, a weather/search radar, smoke and flare dispensers, a flight director system with an auto hover mode, and folding blades for shipboard storage. The RAF made use of this version as the HAR Mk3 and the West German Navy as the Mk 41. Westland exported Sea King variants to India, Norway, Belgium, Pakistan, Australia, and Qatar.

In 1962 the U.S. Marine Corps awarded Sikorsky Aircraft a contract to design and build a new all-weather assault transport helicopter. The S-65, designated the CH-53A Sea Stallion, made its first flight on October 14, 1964, and in 1966 it entered service as the Marine Corps's heavy lift helicopter, as well as replacing the H-3 as the primary USAF SAR helicopter. The Navy also purchased the Sea Stallion to bolster its ASW and SAR operations. Twin T64-GE-6 turbines powered a 79-foot, seven-bladed main rotor that turned above a 73-foot, 4-inch fuselage, designed for emergency water landings. The tailboom supported a tall vertical fin, which in turn supported a horizontal stabilizer attached to the starboard side and a four-bladed tailrotor mounted on the port. Capable of carrying a 14,000-pound load, which might include thirty-eight troops or twenty-four stretchers and three or four medical attendants, the CH-53 "Super Jolly Green Giant" could also accommodate fifty-five passengers with center seating installed. In one emergency situation in Vietnam the CH-53 lifted off with seventy-five Marines aboard. Large rear doors and a ramp facilitated loading either wheeled or palletized cargo.

Designed with a flight direction system and folding blades for shipboard or land-based operations, the big machines soon earned their own distinctive place in aviation history. In August 1970 two CH-53s assigned to the U.S. Air Force Air Rescue and Recovery Service (ARRS), using integral aerial refueling systems, completed a nonstop flight across the Pacific Ocean. The same year the United States used HH-53As in the daring raid on the Son Tay prisoner of war camp, only 23 miles outside Hanoi. Although the pilots and Special Forces troops executed a perfect raid, the NVA had previously moved the prisoners to another location. Variants of the helicopter include the HH-53A/B, HH-53C, CH-53D, RH-53D, CH-53G, and

HH-53H. Sikorsky Aircraft built a total of 412 Sea Stallions with an additional 110 built in the Federal Republic of Germany. Austria, Iran, Japan, and Israel also employed variants of the CH-53. Sikorsky engineers later used the S-65 as a basis for developing the three-engined CH-53E Super SeaStallion/MH-53E Sea Dragon.

The Sikorsky Aircraft S-64, which first flew on May 9, 1962, was the last Sikorsky project in which Igor himself was personally involved. It was a company-funded "flying crane" designed to lift a minimum of 10,000 pounds; Pratt & Whitney T73-P-1 4,500-horse-power engines turned the 79-foot, 6-inch six-bladed main and 15-foot, 4-inch four-bladed conventional tailrotors. Final versions of the helicopter could carry a maximum load of 22,800 pounds. Able to reach a top speed of 111 knots, the large aircraft had a service ceiling of 13,000 feet. Resembling a large insect, the machine sat on an extended stiltlike landing gear that allowed the helicopter to straddle large, palleted cargo loads, or a "people pod" capable of holding forty-five infantrymen or twenty-four litters and fifteen seated passengers. The helicopter could also transport pods containing a mobile headquarters or field hospital. Two pilots sat side by side in the forward flight deck, and a third pilot sat in a lower, rear-facing cockpit that allowed him to fly the aircraft during slingload operations.

It was first delivered to the West German Defense Ministry in 1963, and the U.S. Army ordered six of the machines in June 1963; they eventually bought ninety-seven CH-54 Tarhees between June 1964 and 1972, the last being retired in 1993 from the 113th Aviation Company of the Army National Guard. The CH-54 set several international records, including a time-to-climb record and, in 1965, a world record by lifting off with ninety persons aboard, including eighty-seven combat-loaded paratroopers of the 82d Airborne Division. During the aircraft's service in Vietnam, in addition to a gamut of heavy lift missions in support of U.S. and RVN operations, the "flying crane" recovered 380 downed aircraft. The "Sky Cranes" continued flying in civilian service, fighting forest fires, and in construction, remote oil exploration, and logging operations.

NATO

In Europe during the 1960s and early 1970s, several NATO countries began to produce significant helicopters of their own design. In 1959, Ludwig Bölkow's West German company introduced the

Bo-102 "Heli-Trainer." Designed to train future helicopter pilots in the techniques of hovering and rotary-wing aerodynamics, it had a 40-horsepower ILO piston engine turning a single 21-foot counterbalanced fiberglass rotor blade. Although unable to fly, the little trainer rose to a 2-foot tethered hover, turned about its vertical axis, and allowed prospective pilots to master the difficult art of maintaining a stabilized hover. Bölkow produced at least eighteen of the Bo-102s, which were used by several European countries to train military pilots.

On September 9, 1961, the Bo-103, a flying version of the Bo-102, took to the air. It was basically a larger single-seat version of the Bo-102, with an Augusta GA V four-cylinder piston engine powering the little tubular-framed helicopter. Although it never reached the production stage, the Bo-103 provided the technical basis for the composite rigid rotor system adopted for the very successful Bo-105.

In September 1966 the first Bo-105 prototype, designated the V-1, powered by twin Allison 250 C18 turbines and utilizing a Westland Scout articulated rotor system, experienced ground resonance severe enough to destroy the aircraft. On February 16, 1967, at Ottobrunn, near Munich, a second similarly powered prototype, the V-2, utilizing the composite four-bladed rigid rotor system destined for production machines, made its maiden flight. On December 20, 1967, a third prototype flew with dual 375-horsepower MAN 6200 turboshafts.

The Bölkow Company began the Bo-105 project in 1962 to produce a light, all-weather, general-purpose helicopter capable of carrying a pilot and four to five passengers. The V-2 demonstrated such agility in flight and stability at a hover that the company produced the preproduction Bo-105 V-4 (1969) and V-5 (1970), which became the prototype of the Bo-105C. Additionally, in the spring of 1970, the Messerschmitt-Bölkow-Blohm and Voss group (MBB), which Bölkow's company joined in June 1968, introduced the revolutionary "droop snoot" rotor blade design, which became standard on the Bo-105.

The spacious cabin and high performance of the Bo-105 won rapid acclaim from both military and civilian users. The stubby helicopter reached airspeeds of 130 knots, a service ceiling of 10,000 feet, and, with normal loads, maximum ranges of 310 nautical miles. With engines mounted atop the fuselage, the helicopter featured an oval-shaped cargo compartment/cabin with rear clamshell doors that allow rapid loading and conversion from light cargo to medical evacuation. The unlittered cargo compartment accommodated three troops or two litters. The short tailboom supported a two-bladed tailrotor mounted on the port side and horizontal stabilizers with end-plate vertical fins. The armed version, the PAH-1, carried eight TOW missiles, and more current versions the HOT Anti-Tank Guided Missile (ATGM) system, or a combination of missiles and rockets and machine gun pods. A telescopic day-only sight was installed on the port side of the cockpit. Without armament the aircraft could slingload up to 2,700 pounds. In addition to 299 aircraft ordered for both the West German Army and Navy, the German government also ordered an additional twenty Bo-105s for its *Katastrophenschutz* program, to assist in disaster relief.

In rapid succession the Bo-105 won approval and was certificated by several countries: first in West Germany in 1970, followed by the U.S. Federal Aviation Administration (FAA) in 1971, both the United Kingdom and Canada in 1973, and by the Italian Aeronautical Register in 1974. In 1972 the Bo-105 entered full production with the option of either the Allison 250-C18 or the more powerful C20 engines. Iraq, several Persian Gulf states, Spain, the Philippines, and Indonesia bought military variants of the Bo-105. Additionally, CASA in Spain, NAM in the Philippines, and Nurtanio in Indonesia produced their own Bo-105s under license.

West of the Rhine River, in 1962, the French Army issued a requirement for a French-produced tactical helicopter that could transport twenty soldiers and perform a variety of cargo-carrying duties. The SA 330 Puma, a completely new design, resulted. In 1963, with French government funding, the design process began; the first prototype flew in April 1965. The successful prototype resulted in an order for six preproduction machines. The Puma's twin Bastan VII turbines were mounted on top of the fuselage, leaving the internal fuselage unobstructed for cargo, or for eighteen troops plus two pilots. Replacing the Bastan engines on production models, two Turbomeca Turmo 3C 1,320-horsepower turboshafts drove the 49-foot, 3-inch four-bladed main rotor and five-bladed tailrotor. The Puma exhibited a maximum speed of 150 knots and a range of 340 nautical miles.

Impressed by the Puma, both the French Army and the UK Royal Air Force placed substantial orders. On February 22, 1967, Sud-Aviation entered into an agreement with Westland Helicopters to coproduce the Puma. Under the contract the companies jointly manufactured the SA 330 Puma, the SA 341 Gazelle, and the WG Lynx. Thus all Pumas were partially manufactured in England; Westland assembled completely the forty-eight HC1 Pumas ordered by the British military, and the first was delivered on July 30, 1968. The agreement that also allowed the French to build forty units of the Westland Lynx remained in force until 1988. On January 1, 1970, Sud-Aviation, Nord Aviation, and SEREB merged to form the Société Nationale Industrielle Aerospatiale, which honored all contracts previously signed with other manufacturers.

Several other countries purchased military variants of the Puma, including Chile, Indonesia, Morocco, South Africa, and Spain. Romania built the SA 330 under license as the IAR-330 and produced about ninety military and civilian models that were used by Romanian military and commercial operators, as well as the militaries of Ethiopia and Guinea. Established in 1963 as the major maintenance support for the South African Air Force (SAAF), Atlas Aircraft Company Ltd. completed a major modification of the SAAF Pumas, first known as the Gemsbok, then the Oryx. Atlas installed the more powerful Makila 1A turboshafts, new gearboxes, and extensive upgrades of weapon systems and avionics, including a nose-mounted radar. Atlas subsequently developed a gunship version of the Puma designated the ZTP-1 Oryx, on which the company installed external stub wings, with FFAR and machine gun pods, and a TC-20 20-mm cannon mounted under the fuselage. Bristow Helicopters also ordered a number of Pumas for off-shore work in the rough weather of the North Sea.

In response to a French Army requirement for a new observation and reconnaissance helicopter to replace the Alouette, Sud-Aviation introduced the SA 340 Gazelle prototype, which took flight on April 7, 1967. An all-metal five-seat helicopter, the SA 340 initially incorporated an Alouette II tailrotor but was subsequently fitted with a "Fenestron," a multiblade antitorque device housed in a circular shroud within the tailfin itself. On April 12, 1968, the SA 341 Gazelle appeared with the Fenestron at the end of a 31-foot, 3-inch fuselage/tailboom and a 34-foot, 5-inch rigid main rotor system designed by Messerschmitt-Bölkow-Blohm, based on the German MBB Bo 105. Sud-Aviation flew four production prototypes of the distinctive SA 341 and included the aircraft in the 1967 joint production agreement with Westland. Powered by a Turbomeca/Rolls-Royce Astazou 111N 592-horsepower turboshaft, the Gazelle received initial orders from both the French and UK military establishments. On August 6, 1971, the first production Gazelle rolled off the assembly line at Toulouse, France.

Although the Gazelle was designed primarily for aerial observation, its agility, initial top speed of 143 knots, and range of 360 nautical miles made the helicopter much more versatile. The eleven variants of the aircraft performed a myriad of roles, including airborne Forward Air Control, air ambulance, liaison, and command and control; equipped with a Ferranti AF 532 stabilized sight and HOT antitank missiles, it became a mainstay among European antiarmor helicopters. In May 1973, Aerospatiale introduced the SA 342 Gazelle with a more powerful 870-horsepower turbine engine. Westland produced a total of 262 Gazelles of all models, most delivered to the UK military. Egypt acquired almost 100 Gazelles, and Iraq, Libya, and Kuwait procured several Gazelles for their military. Under license in Yugoslavia, Soko built 132 Gazelles, designated the SA 321H Partizan, as well as a number of the SA 342L with an improved Fenestron. Commercial operators throughout the world bought several civilian versions of the Gazelle.

On June 2, 1972, Aerospatiale introduced the SA 360 Dauphin (Dolphin) to fill the gap between the Gazelle and Puma. Derived from the Alouette, the first variant of the Dauphin had the Alouette rotor system, the Fenestron of the Gazelle, and a single Turbomeca Astazou XVI 890-horsepower turboshaft. Neither the military nor civilian versions of the SA 360 met with much success, and in 1973 engineers began a redesign program. First flown in 1975, the revamped SA 365 Dauphin came in either single- or twin-engine models with the turbomeca Astazou XVIIA 1,050-horsepower gas turbine engine. Capable of carrying six to ten passengers in its 36-foot fuselage, the Dauphine featured Aerospatiale's new "Starflex" 37-foot, 8-inch rotor system, which incorporated a central composite head along with maintenance-free elastomeric ball joints to reduce both wear and maintenance costs. In its initial configuration, the Dauphin, with a demonstrated airspeed of 170 knots in level flight, set several speed records and received several improvements as orders for the aircraft increased. Not until July 1978 did the SA 365C receive its French commercial certificate, resulting in several orders for civilian models, but military orders remained scarce.

In the United Kingdom, Westland continued to produce helicopters of its own design and those of other companies under license. On October 28, 1962, the company introduced the HAS Mk 1 Wasp, a naval version of the Mk 1 Scout. The Wasp measured 40 feet, 4 inches in length, with a main rotor diameter of 32 feet, 3 inches. A single Rolls-Royce Nimbus 103 1,050-horsepower turbine, derated to 710 horsepower, drove the five-place, 5,500-pound helicopter to a maximum speed of 104 knots. The Royal Navy stationed the aircraft aboard frigates and equipped them for ship attack, ASW, and SAR.

Weapon loads included two Mk 44 or Mk 46 homing torpedoes, or 540 pounds of bombs or depth charges. Several Wasps carried Nord AS 11 or AS 12 antiship missiles. Westland constructed ninety-eight machines for the RN, Indonesia, and Malaysia.

On March 9, 1965, Westland produced the first of 149 AH Mk 1/HT Mk 2 Sioux helicopters. Basically a Bell Model 47G three-seat light utility helicopter built under license from Bell and Augusta, it had an Avco Lycoming TVO-435-A1A 260-horsepower turbo-charged piston engine. The Sioux, manufactured with the Bell semi-rigid rotor system, flew at 90 knots and was employed by the British military as a general purpose and training helicopter.

In August 1965, Westland began a program to build three new Wessex HAS Mk 3 helicopters and convert forty-three RN Mk 1s to the Mk 3 standard. Although the improved version made use of the basic WS-58 airframe, the Mk 3 incorporated a much refined avionics and automatic flight control system (AFCS), along with an updated ASW dipping sonar and homing torpedoes.

The last half of the 1960s provided additional milestones for Westland. On October 1, 1966, the company reorganized under the name Westland Helicopters Ltd. On the 8th of the previous month, under a licensing agreement, Westland began manufacturing the Sikorsky SH-3D Sea King as the HAS Mk 1, 2, 5, and 6, and the HAR Mk 3. Westland's modifications included equipment upgrades to improve the Sea King's utilization as an ASW platform and subsequent adaptations for SAR, commando insertions, and airborne electronic warfare (AEW). Westland's Sea King rotor diameter measured 62 feet, and the helicopter 72 feet, 8 inches in length. Two Rolls-Royce Gnome 1400 1,500-horsepower turboshafts powered the aircraft to a speed of 114 knots; it had a maximum gross weight of 21,400 pounds. The ASW version carried a pilot and three system technicians, while the Commando version carried two pilots and up to twenty-eight troops to a maximum range of 600 nautical miles without refueling. Westland built a total of 330 of all versions.

On March 21, 1971, the prototype WG 13 Lynx, the third of the trio of helicopters involved in the Anglo/Franco collaboration, took to the air, with Westland holding the design authority for the aircraft. Although the first Lynx was destined for the British Army, Westland engineers intended the 42-foot, semi-rigid rotor system to offer speed and excellent maneuverability for shipboard operations. Westland initially equipped the Lynx with two Bristol-Sidney BS 360 turbines, but two Rolls Royce Gem 2 900-horsepower turboshafts replaced the BS engines when they exhibited excessive vibration and oil consumption. The new engines provided power to fly the Lynx at 140 knots at a maximum gross weight of 10,000 pounds. The Lynx became the first British helicopter to complete a roll.

The utility version of the Lynx became operational with the British Army in the mid-1970s, and, with several upgrades such as the AH Mk 7 and the AH Mk 9, the 150 Army versions produced by Westland have remained in service into the twenty-first century. The utility version of the Lynx carried two pilots and ten combat-loaded troops, or conducted resupply or air ambulance missions. The armed version carried a pilot and copilot/gunner and was armed with eight TOW antitank missiles, or a combination of 20-mm cannon and 7.62-mm miniguns. The armed Lynx provided escort for troop-carrying helicopters, or filled the aerial reconnaissance and tank killer roles.

On May 25, 1976, the HAS Mk 1 Lynx began flying with the RN. Designed specifically to operate on small ships, the naval Lynx featured both a folding tailboom and main rotor blades, as well as a high-energy-absorbing wheeled landing gear with a deck lock. In addition to the wheeled landing gear as opposed to skids, the Navy Lynx differed from the Army version in that the engines produced 1,120 horsepower each, permitting an increase of 1,300 pounds in gross weight, but additional equipment reduced airspeed to 125 knots. ASW versions carried two technical operators in addition to the pilots and were armed with Mk 44, Mk 46, or Sting Ray torpedoes. Westland built 250 HAS Mk 1, HAS Mk 3, and HAS Mk 7 versions of the naval Lynx.

In Italy, Giovanni Agusta, in 1907, formed the Costruzioni Aeronautiche, which produced several types of aircraft in factories in Cascina Costa and Milan. For a time after World War II the company turned away from aircraft production, but in 1952, Domenico Agusta realized the future of helicopters and concluded an agreement to produce Bell helicopters under license. Since that time Agusta has consolidated C.A.G. Agusta S.p.A., Bredanardi S.p.A., SIAI Marchetti S.p.A., Caproni Vizzola S.p.A., and Industria Aeronautica Dionale S.p.A. into the Grupo Agusta, a significant European aerospace manufacturer. In June 1980 the Agusta Group (ITA) entered into a partnership with GKN-Westland in European Helicopter Industries Ltd. In May 1954, Agusta completed the first Bell 47G built under its license, with other various models following. The Italian Air Force, Navy, and Carabinieri received about one-third of more than 700 Model 47s built.

From the late 1950s through the 1970s, Agusta experimented with several company-designed helicopters, but none met with real success. In 1958 a prototype four-engined AZ 8 flew for a short period; it led to another prototype in 1964, the AZ 101. Three Rolls-Royce-Bristol Gnome turbines drove a five-bladed rotor system. Agusta hoped to compete with the Frelon and Sikorsky S-61 with the thirty-eight passenger AZ 101G, but the helicopter failed expectations and no Italian military orders materialized. Agusta, therefore, entered into license agreements with Sikorsky and Boeing-Vertol to build helicopters that would meet the needs of the Aeronautica Militare Italiana.

In an attempt to provide customers with an aircraft that would offer greater seating capacity, and eventually replace the Bell 47, Agusta experimented with several machines. Ironically, the first was a redesigned Bell 48 (YH-12B) prototype that had first flown in 1946. Agusta engineers combined a new nine-seat fuselage with the engine and rotor system of the Bell 48. The resulting machine became the first helicopter to earn a type certificate in Italy. A commercial operator bought two A 102s, but turbine-powered helicopters doomed the A 102.

Agusta responded with several experimental turbine engine models. A prototype of the Bell 47G powered by a Turbomeca Astazou turboshaft engine did not live up to expectations and was canceled. Two small prototypes followed, the A 103 and A 104, both powered by a 140-horsepower MV-Agusta piston engine. On the follow-on A 104BT, Agusta engineers replaced the piston engine with a light turbine engine, but this variant also proved unsatisfactory. In 1965, the final variant, the A 105, appeared with a more powerful Turbomeca-Agusta turboshaft engine, but again, the aircraft did not sell; neither did the four-seat A 105B.

Customers wanted helicopters with larger seating capacity, and Agusta decided to produce the Bell Model 206 Jet Ranger under license rather than pursuing the expensive option of developing and producing a helicopter of its own design. The Jet Ranger proved to be a highly lucrative program for Agusta, and eventually the company produced more than 900 variants of the Bell 206. Agusta also concluded agreements to build substantial numbers of the Bell 204, 205, 212, and subsequent models for both civilian and military operators, including the specialized AB 212ASW variant for the Italian Navy.

In 1967, Agusta opened another profitable line of helicopters by producing Sikorsky helicopters under a separate license agreement. The Italian Navy ordered twenty-six of the Agusta-manufactured Sikorsky Sea King (AS-61A-4) ASH-3Ds equipped with the

AN/APN-195 radar installed in a nose radome and the APQ-706 Marte missile guidance radar mounted in a pod under the fuselage.

Agusta also produced two ASH-3D/TS VIP versions for the Italian government and the papacy. Agusta built a total of eighty Sea Kings, with sales to Libya, Peru, Iran, and Saudi Arabia.

Additionally, Agusta manufactured the Sikorsky HH-3F Pelican in search and rescue versions for the Italian military. The 15th Stormo received twenty of the aircraft and stationed them at bases along the Italian coast. SIAI also produced the HH-3F and the AB 205, AB 212, and, in the 1980s, the AB 412. Agusta, through an association with Elicotteri Meridionali, built more than 160 Boeing CH-47C Chinooks under license for the militaries of Italy, Egypt, Libya, Morocco, Greece, and Iran.

BredaNardi, a member of the Agusta Group, signed license agreements to produce helicopters designed by another U.S. manufacturer. Initially the company produced the piston-engined Hughes 300C at its Milan plant, with the Greek Air Force ordering twenty machines for pilot training. The company also built more than 150 of the turbine-powered 500 series for both military and civilian customers. Recently, the BredaNardi factory became the Monteprandone RKS division of Agusta.

Agusta's agreements with Bell, Sikorsky, and Hughes restricted the company from sales in North America, and government demands for an Italian designed and built helicopter initiated an autonomous program in 1964. The A 109 emerged as a streamlined pod and boom-type helicopter capable of carrying six passengers and two pilots in a single cabin. Planned as a single-engined aircraft with an Astazou 12 turboshaft, the A 109 soon outgrew its single engine. Along with a separate passenger compartment and a two-place flight deck, Agusta engineers redesigned the fuselage to accommodate two Allison C14 turboshafts. Agusta constructed five prototype airframes in 1970, with one flying the following summer. In 1975 the A 109 received its type certification from the Italian government and went into full production.

THE MIDDLE EAST

The 1960s and 1970s also saw helicopters become an integral weapon system in Middle Eastern conflicts. In April and May of 1967, Egyptian president Gamal Abdel Nasser, responding to false

Soviet reports that Israel was planning an attack on Syria, mobilized his military in preparation to invade Israel. On June 5, Israel launched a surprise attack that destroyed most of the Egyptian Air Force. During the next six days the Israeli Defense Force (IDF) defeated in detail Egypt, Jordan, and Syria's military forces. IDF Sikorsky and Bell helicopters ferried troops and supplies, evacuated wounded, and rescued downed pilots from far behind enemy lines. Israeli paratroopers inserted by assault helicopters captured the southern half of the strategic Golan Heights.

Unfortunately, on June 8, 1967, Israeli helicopters participated in what Israel alleged to be a mistaken attack on the USS *Liberty*, a U.S. Navy intelligence-gathering ship operating in international waters. IDF jet fighters and motor torpedo boats swept in firing rockets and cannon at the ship in an attack lasting more than two hours. Troop-carrying helicopters then appeared and machine-gunned the life boats. In all the United States suffered 54 sailors killed and another 171 wounded.

During the "War of Attrition" between Egypt and Israel (1967–1970) Israel employed helicopters on numerous occasions. When Arab terrorists infiltrated into Israel from the Jordan Valley, the IDF retaliated with swift strikes using heliborne infantry supported by rocket-firing helicopters and jet fighters. In one instance IDF commandos, utilizing CH-53 Super Jolly Green Giants, swept in deep behind the Suez Canal and captured an Egyptian P-12 radar system's acquisition and command trailer. The Israelis quickly rigged the trailer for a slingload, and all the IDF forces returned safely with the purloined radar equipment.

On October 6, 1973, Egypt and Syria initiated a surprise attack against Israel to reclaim their land, and reputation, lost to the IDF in 1967. Initially surprised and forced to retreat, the IDF quickly reacted and began a series of counterattacks that drove the Arab forces back across the 1967 armistice line. During the Yom Kippur War, or Ramadan War as it is known in Islamic countries, assault helicopters played a major role in Israel's triumph by rushing troops and supplies to critical points on the battlefield, but the IDF lost large numbers of tanks because of a lack of attack helicopter support. After several engagements, wires from Soviet-manufactured wire-guided ATGMs crisscrossed Israeli tanks "like a spider web," prodding the Israeli government to order numerous Bell AH-1 and Hughes OH-6 TOW-firing helicopters to provide greater standoff range from Arab armored units. The shah of Iran also realized the importance of both transport and tank-killing helicopters in the desert and began an extended purchase of Bell Model 214A and twin-engined NAH-1J Cobras.

WARSAW PACT

In 1962, WSK-Swidnik in Poland began producing the Mi-2 Hoplite under license, although the plant did not reach full production until 1965. With two Isotov GTD-350 turboshafts mounted above the cabin, the Mi-2 produced 40 percent more power, at less than half the weight, effectively doubling the lifting capability of the Mi-1. An all-metal three-bladed rotor system, measuring 47 feet, 9 inches in diameter, provided lift for the single-pilot, eight-passenger utility helicopter. Of a modified pod and boom design, the Mi-2 attained a maximum airspeed of 110 knots and hauled 3,000 pounds of internal cargo, or a 1,750-pound slingload. External pylons provided mounting points for a combination of machine guns and AT-2 Sagger ATGMs. The Mil Company reported the maximum service ceiling at 13,000 feet.

In the late 1950s, to meet several requirements of the Soviet military, Mikhail Leontyevich Mil began designing and producing large heavy-lift helicopters. On June 5, 1957, Mil Design Bureau test pilot Rafail Kaprelian piloted the Mi-6 Hook on its maiden flight. The first of five prototypes, the Mi-6 was the world's largest helicopter to that time, and the Soviet Union's first production turbine-powered helicopter. The fuselage measured 134.5 feet, and the five-bladed, all-metal main rotor system 114.8 feet in diameter, dwarfing all previous helicopters. Two removable wings, with a span of 50 feet, 2 inches, provided additional lift in forward flight. The groundcrew usually removed the wings for large slingload operations, as they just increased drag.

Mil utilized plastic-impregnated wood, reinforced with steel spars, to manufacture the four-bladed tailrotor. Production models carried an alcohol-based deicing system. Equipped with two Soloviev D-25V turbine engines, producing 5,500 horsepower each, the huge machine could carry up to ninety troops and lift off at a maximum gross weight of 93,500 pounds, which translated into a payload of 26,400 pounds of internal cargo, or a 17,500-pound slingload. As an air ambulance the Hook could accommodate forty-one stretchers and two medical attendants. Two hydraulically operated rear-opening clamshell doors made possible the loading of large cargo into the fuselage, which included a winch with a capacity of 1,750 pounds. A pilot, copilot, navigator, flight engineer, and radio operator composed the crew. In 1961 the Mi-6 became not only the largest but also the fastest helicopter in the world by reaching 300 kph (162 knots), thereby winning the Igor Sikorsky International trophy. The Hook had a range of 385 statute miles and a service ceiling of 14,750 feet. For ferry flights the Mi-6 could carry four auxiliary fuel tanks, two installed internally and two mounted above the main wheels of the fixed tricycle landing gear.

The standard avionics package provided the Hook with day/night, all-weather capabilities. The helicopter usually carried a ground power unit (GPU) on a trolley for engine starting and ground operations. A single DShK 12.7-mm heavy machine gun could be installed in a flex mount in the nose for self-protection. From the time the Mi-6 entered production in late 1960, until production ended in 1981, Mil produced 860 aircraft. Aeroflot used the large helicopters in trackless Siberia, while the Soviet military utilized the heavy-lift capabilities of the machine to transport rocket launchers and other heavy weapons. The Mi-6VKP, Hook "B," hauled a huge array of radios and functioned as an airborne command post, which eventually led to the Mi-22C "command post." The Hook, in various modifications, provided forward refueling points for both vehicles and aircraft, fought forest fires, and recovered *Soyuz* space capsules. The USSR produced 860 Hooks and exported the machine to several countries, including North Vietnam, which used the Hook to carry fighter aircraft from airfields to remote locations to protect the jets from U.S. air strikes. In July 2002 an Mi-6 crashed, killing all twenty-one on board, and the Russian Ministry of Transportation halted the use of the aging Mi-6.

Continuing to focus on heavy-lift helicopters, the Mil Bureau introduced the Mi-10 Harke. In July 1961 the "flying crane" modification of the Mi-6 appeared at the Soviet Tushino Air Show; it flew at the Paris Air Show in 1965. The Mi-10 shared engines, transmissions, rotor system, hydraulics, and many other parts with the Mi-6, reducing maintenance costs for both aircraft. The Mi-10, however, featured a much modified fuselage designed for transporting large external loads weighing up to 36,300 pounds. Mil designed the machine with an extended, wide, four-legged dual-wheeled landing gear to straddle large, cumbersome loads. For whatever reason, the bureau fitted the aircraft with right-side struts a foot shorter than those on the left, causing the machine to cant slightly to the right when parked.

Several other modifications differentiated the Harke from the Hook. Hydraulic arms fitted to the fuselage underside held specially designed wheeled pallets in place during flight. It was intended primarily as a cargo helicopter, and there were no provisions to install removable wings; the shortened, 107-foot, 10-inch fuselage, however, could accommodate up to twenty-eight passengers in addition to the three-man crew. The first models included a pod, or "dustbin," under the nose by which the crew could observe external loads, but the pod was removed from later models and the crew monitored their loads via closed-circuit television. Slower than the Hook, the Mi-10 reached a top speed of 120 knots and a service ceiling of 9,850 feet.

The Mi-10K variant appeared in 1966. Differing slightly from previous models, the "K" offered greater lift capacity, shorter landing gear struts, and a gondola under the forward fuselage. The gondola, reflecting that on the Sikorsky CH-54, housed a full set of flight controls, allowing a copilot to position unwieldy payloads accurately. Certain specialized variants included the Mi-10P electronic countermeasures (ECM) helicopter. The Soviet Union produced only fifty-five Mi-10s, and none are known to have been exported.

In 1965 the Mil bureau began work on the largest helicopter ever built, the Mi-12 Homer. Designated the V-12 by the Soviets, the aircraft, comparable in size to a Boeing 727, first took flight in 1968. The next year a prototype lifted a 72,000-pound payload to 7,900 feet, setting a new weight-to-altitude record. Capable of carrying 120 passengers, the Homer made its first public appearance at the 1971 Paris Air Show.

Designed by Mil, who died in 1972 and was replaced by Marat Tischenko, the Mi-12 encountered numerous developmental problems. Of a side-by-side rotor design, the Homer incorporated the engines and transmissions of the Mi-6. At the end of two reverse-tapered wings (wider at the tip than at the root), two Soloviev 5,500-horsepower turboshafts drove a five-bladed rotor system. The Mi-12 sat on a dual-wheeled tricycle landing gear, and large clamshell doors at the rear of the fuselage allowed loading of oversize cargo. Requiring a crew of six, the Homer never fully met its design specifications, and the program was canceled. One prototype remains on static display in Russia.

In May 1960, Mil conceived a machine to replace the piston-engined Mi-4 Hound. On June 9, 1961, the first Mi-8 Hip prototype, with a single AI-24V turboshaft and four-bladed main rotor system, lifted off for its maiden flight. On September 17, 1962, the

Hip B, modified with two TV2-117 1,482-horsepower turboshafts mounted atop the fuselage, and a five-bladed main rotor system measuring 70 feet in diameter, took flight. The Mi-8 went into full production in 1965, and by 2000 fifty-four countries operated the more than 10,000 Mi-8s manufactured by the Rostov and Kazan production facilities in Russia and by foreign licensees. Designed as a medium-lift transport helicopter, the Hip, in its many variants, fulfilled a miscellany of mission requirements, including troop and cargo transportation, air ambulance, attack helicopter, airborne command post, fire fighter, and civilian carrier.

Constructed of light alloys, the Hip featured a "bus-shaped" fuselage with a rounded nose and glassed-in cockpit that accommodated a pilot, copilot, and flight engineer. The cabin housed twenty-four passengers, 8,800 pounds of cargo, or twelve stretchers. A large sliding door on the forward port side and rear-opening clamshell doors simplified loading large cargo. Removable interior seats and an internal winch capable of lifting 350 pounds that doubled as a rescue hoist facilitated cargo handling. Additionally, Mil equipped the aircraft with a cargo hook capable of carrying slingloads up to 6,500 pounds. A long tailboom extended from the upper portion of the fuselage and swept up to a tapered vertical fin that housed the gearbox and tailrotor, attached to the left side (right on the export versions).

External racks attached along the center of the 61-foot fuselage were designed to hold auxiliary fuel pods or weapons systems. Variants of the Hip carried a combination of 57-mm or 80-mm rockets, AT-2 Swatter or AT-3 Sagger ATGMs, 12.7- or 23-mm gun pods, or either 4 × 500-pound or 2 × 1,000-pound bombs. In 1967, Mil introduced the Hip E and F ground support helicopters, each mounting a flexible 12.7-mm heavy machine gun under the nose and carrying 192 57-mm rockets. Combat troops could also fire their individual weapons from the windows of the helicopter. In later models Mil installed the upgraded Isotov TV2-117A engines, which produced 1,700 horsepower each. Generally a Hip cruised at 122 knots, had a service ceiling of 14,700 feet, and hovered Out of Ground Effect (OGE) at 2,600 feet. All Mi-8s rested on a fixed tricycle landing gear, with dual wheels at the nose. Total production estimates ran as high as 15,000 units of the Mi-8 and its export version, the Mi-17.

In September 1969 the Mil Design Bureau modified the Hip into the Mi-14 Haze for naval applications, mainly for shore-based ASW operations. The Mi-14 received a boat-hulled lower fuselage with pontoons on either side and a retractable landing gear. A radar dome under the nose and an internal weapons bay differentiated the Mi-14 from the Mi-8. Both the Mi-8 and Mi-14 carried infrared jammers and flare/chaff dispensers.

Another modification allowed the TV2-117TG engines to operate on both liquefied petroleum gas (LPG) and standard jet fuel (a type of kerosene). Large external tanks held the LPG under low pressure, and the pilots switched the engines to regular fuel for takeoffs and landings. The LPG tanks reduced the helicopter's payload by 220 to 330 pounds but extended its useful range by several miles.

In July 1961 the Kamov Ka-20 Harp appeared for the first time during the Soviet Aviation Day celebration. The Ka-20 followed the traditional N. I. Kamov coaxial rotor design, with two three-bladed, counter-rotating main rotors. A large bulge under the forward fuselage and a fairing under the tailboom indicated that the Ka-20 was an ASW helicopter of some type. It was much larger than the Ka-18, with dual turbine engines mounted over the cabin body. Two fixed machine guns protruded from the nose, and two air to surface missiles (ASM), probably dummies, hung from external racks on each side of the helicopter. The new craft proved to be an advanced prototype of the Ka-25 Hormone, which became operational with the Soviet Navy in 1965.

Only slightly different from its predecessor, the Kamov Bureau (OKB) produced the Ka-25 Hormone in three major variants, all powered by two Glushenko GTD-3 900-horsepower turboshafts, as well as featuring a coaxial rotor system and the compact body most effective in shipboard operations. OKB designers included folding main rotor blades, reducing the Ka-25's stowed length to only 36 feet. The Hormone A, designed specifically to destroy nuclear-powered submarines, carried two pilots and three sensor technicians to operate ASW equipment, which included a search radar installed in a large, chin-mounted radome, a towed magnetic anomaly detector (MAD), and a dipping sonar housed in a compartment at the rear of the cabin. Some models featured ventral weapons bays by which the Ka-25 could be armed with torpedoes, depth charges (including nuclear), or air to surface missiles. Later improvements to the Hormone included an automatic pilot, state-of-the-art avionics, and a vastly improved over-the-water navigation system for precise targeting of submarines and surface vessels. By 1968 the Hormone A operated from cruisers of the *Kresta* and *Kara* classes, *Moskva* and *Leningrad* carrier/cruisers, and *Kiev* and *Minsk* ASW carriers. The smaller cruisers carried only two Ka-25s, while as many as eighteen Hormones operated off the larger ASW carriers.

The second variant carried search radar and other electro-optic target acquisition equipment for over-the-horizon guidance of surface-to-surface missiles and naval rifles. A larger, and more spherical, chin radome, a cylindrical radome mounted at the aft end of the fuselage, and no ventral bays differentiated the Ka-25 B from other models. The Hormone reached an airspeed of 105 knots, a range of 350 nautical miles, and a service ceiling of 11,000 feet.

In a later variant, the Hormone C, OKB engineers removed the ASW and targeting instrumentation and equipped the helicopter for SAR missions. Without the electronic gear the large cabin also offered a secondary capability of transporting twelve naval infantry troops. Two other versions appeared in limited numbers, the Ka-25 BT mine countermeasures helicopter and the Ka-25 K civilian helicopter, which had a small cockpit under the forward fuselage for slingload operations. In the event of ditching, most models of the Hormone were fitted with inflatable pontoons on each of the four landing gear struts. Between 1965 and 1975 the USSR produced more than 460 Ka-25s, with export models sent to Syria, India, Bulgaria, North Vietnam, and Yugoslavia.

In January 1964 the Soviet government issued a directive for both an agricultural and a passenger transport helicopter. On August 18 of the next year, the Kamov Bureau responded with the prototype Ka-26 Hoodlum, a small, multipurpose helicopter. Kamov deputy chief designer M. A. Kupfer and project leading engineer Y. I. Petrukhin, of course, specified the installation of OKB's typical coaxial rotor system atop the boxlike fuselage of the Ka-26. The craft incorporated a conventional Kamov tail assembly of twin vertical fins with rudders and an interconnecting horizontal stabilizer, which included the elevator. The helicopter sat on a four-wheeled fixed landing gear. Initial manufacture began in 1967 at the Kumertau Aircraft Plant. Placed in full production in 1970, the multirole helicopter was powered by two 325-horsepower Vedeneyev M-14V-26 nine-cylinder radial engines. Flown by either one or two pilots, the Hoodlum hauled loads of 2,000 pounds up to altitudes of 9,850 feet, at speeds of 95 knots. OKB's modular design and interchangeable cargo containers permitted rapid conversion of the Hoodlum from a passenger carrier to an air ambulance, crop duster, or "flying crane." Kamov OKB first used composite materials for the engine cowlings, rotor blades, and chemical hoppers, which increased the useful life of Ka-26 subassemblies. The composite rotor blades, for example, demonstrated a service life of 5,000 hours, compared with the 600 to 800 hours of conventional metal blades of the era. With

OKB production methods patented in five Western nations, the Hoodlum became the only Soviet helicopter certificated under U.S. federal aviation regulations (FAR). The Soviet Union produced at least 850 Ka-26s and exported the machine worldwide.

In 1970, Kamov created his first designs for the next generation of Soviet naval helicopters. He envisioned a machine similar in size to the Ka-25 but with an updated avionics and weapons package. On December 24, 1973, OKB's chief pilot, Y. I. Laryushin, lifted the prototype Ka-252 off for its first flight. Unfortunately, the aircraft's designer did not see his project completed; Kamov died on November 24. Sergei Viktorovich Mikheyev then assumed the directorship of OKB, which in 1974 was renamed in honor of Nikolai Il'yich Kamov.

Developed from the Ka-252 prototype, the Ka-27 Helix produced more power from the two Isotov TV3-117 2,225-horsepower turboshafts mounted under the three-bladed coaxial rotor system. Only slightly longer than the Ka-25, the Helix had a redesigned tail assembly. With a maximum gross weight of 24,250 pounds, the Ka-27 rotor diameter measured 52 feet, 2 inches and the aircraft 37 feet, 1 inch in length. Capable of 140 knots, the Helix had a maximum range of 432 nautical miles and a service ceiling of 19,700 feet. Designed specifically for the Soviet Navy as an ASW hunter/killer, the Helix did not become operational until 1978.

CONCLUSION

During the Vietnam War the success of U.S. helicopters radicalized modern warfare. Generals and admirals the world over realized that the helicopter had become an integral, and indispensable, element of war planning. Boeing CH-47 Chinooks, Sikorsky CH-54 Tarhees (Skycranes), CH-3 Sea Knights, and CH-53 Sea Stallions magnified the combat versatility of units in Vietnam by lifting towed 155-mm and 8-inch artillery pieces to remote firebases that provided fire support for infantry units moving into contact. Helicopters inserted large numbers of infantrymen and Marines onto precarious mountaintop LZs and, employing a long rope ladder that dangled down through the trees, delivered Rangers and Special Forces into triple canopy jungles. Outlying firebases and Special Forces camps relied on frequent resupply from U.S. workhorse helicopters. Airmobile units used helicopters to preposition fuel and ammunition for future operations. Forward Area Refueling Points (FARPs) reduced the turnaround time for helicopters to return to their missions. By sling-loading downed aircraft back to repair facilities the CH-37, CH-47, CH-53, and CH-54 salvaged aircraft that would have been a total loss in other circumstances.

Marine helicopter operations followed much the same pattern as the U.S. Army, with a few innovations. Despite visibility limitations, Marine helicopter crewmen conducted several early night operations. In August 1965 they flew the first night assault in Vietnam, using CH-34s to insert Marine grunts into Elephant Valley northwest of Da Nang. Marine medevacs also extracted several severely wounded men during night operations. Radar operators guided the pilots to the wounded while other aircraft dropped flares to illuminate LZs long enough for the medevac to evacuate casualties.

Radar guidance allowed the Marines to resupply units under inclement weather conditions. During the siege of Khe Sahn and operations in the A Shau Valley in 1969, CH-46s and CH-53s made instrument climbs through the overcast at Quang Tri and Da Nang, and with radar directions flew to the beleaguered Marines. When an opening in the clouds appeared the pilots spiraled down to drop their external loads of water, rations, ammunition, and medical supplies into small LZs hacked into the cloud-covered rain forests.

From administrative assignments ("ash and trash" missions) to combat and service operations, helicopters changed U.S. military doctrine forever. Used properly, helicopters proved much less fragile than some doomsayers had predicted, flying thousands of hours per aircraft lost or damaged. By the end of 1972, U.S. helicopter losses neared 4,500, but most were lost to operational accidents and mechanical failures. Many of those listed as destroyed in combat were lost to mortar and rocket fire while the aircraft sat on the ground, partially protected by sandbag revetments.

During the Vietnam War, medevac and SAR pilots flew their aircraft into every conceivable situation: firefights, dense jungles, remote mountaintops, rice paddies, river bottoms, steep ridges, even to the outskirts of Hanoi itself, considered the most defended city in the world. For example, the USAF Aerospace Rescue and Recovery Service, flying a variety of helicopters, rescued 3,883 U.S. servicemen in SEA who might otherwise have been killed or captured. U.S. helicopters also evacuated hundreds of Americans and South Vietnamese during the fall of Saigon in April 1975.

From the late 1950s through the mid-1970s helicopters proved their adaptability and capability to operate in searing desert heat,

Arctic cold, and all environmental conditions in between. Flying a miscellany of missions including military assault and transport, SAR, vertical replenishment of both military and commercial shipping, minesweeping, advanced early warning, and various humanitarian aid missions such as air ambulance and disaster relief, helicopter pilots aptly demonstrated that rotorcraft are much more versatile than conventional airplanes. Flying these missions in all weather conditions, aviators permanently established the helicopter's reputation as a machine that could accomplish any assigned task. As to air ambulance and rescue, Igor Sikorsky, in 1969, stated: "For me the greatest source of comfort and satisfaction is the fact that our helicopters have saved . . . over fifty thousand lives and still continue with their rescue missions. I consider this to be the most glorious page in the history of aviation." Both Larry Bell and Sikorsky believed that of the "great varieties of service rendered by the helicopters . . . the most important [have been] the saving of many thousands of lives" (Keogan 2003, 44).

REFERENCES

Ahnstrom, D. N. *The Complete Book of Helicopters.* New York: World Publishing Company, 1971.

Aircraft of the World: The Complete Guide. NP/USA: International Masters Publishers AB, licensed to IMP, 1996.

Anderton, David, and Jay Miller. *Boeing Helicopter: The CH-47.* Arlington, TX: Aerofax, 1989.

Apostolo, Giorgio. *The Illustrated Encyclopedia of Helicopters.* New York: Bonanza Books, 1984.

Bell Helicopter-Textron, http://www.bellhelicopter.textron.com. Accessed June 2002.

Boeing Company, http://www.boeing.com. Accessed January 2003.

British Army Air Corps Association, http://www.aacn.org.uk. Accessed May 2002.

British Army Air Corps Historical Flight, http://www.rdg.ac.uk. Accessed May 2002.

British Army Air Corps Museum, http://www.flying-museum.org.uk. Accessed May 2002.

British Army Official Site, http://www.army.mod.uk. Accessed May 2002.

Brown, David A. *The Bell Helicopter Textron Story: Changing the Way the World Flies.* Arlington, TX: Aerofax, 1995.

Carlson, Ted. "Marine Twin Hueys." *World Airpower Journal* 42 (autumn/fall 2000), pp. 134–143.

Chant, Christopher. *Fighting Helicopters of the 20th Century.* Christchurch, Dorset, England: Graham Beehag Books, 1996.

Coleman, J. D. *Pleiku: The Dawn of Helicopter Warfare in Vietnam.* New York: St. Martin's Press, 1988.

Cook, John L. *Rescue under Fire.* Atglen, PA: Schiffer Books, 1998.

Cowin, Hugh W. *Military Helicopters.* New York: Gallery Books, imprint of W. H. Smith Publishers, 1984.

Delear, Frank J. *Igor Sikorsky: His Three Careers in Aviation.* New York: Dodd, Mead, & Company, 1969.

Donald, David, ed. *The Complete Encyclopedia of World Aircraft.* New York: Barnes and Noble Books, 1997.

Dorr, Robert F., and Chris Bishop. *Vietnam Air War Debrief.* London: Aerospace Publishing, 1996.

Dowling, John. *RAF Helicopters: The First 20 Years.* London: Her Majesty's Stationery Office, 1992.

Everett-Heath, John. *Soviet Helicopters.* London: Jane's Publishing Company, 1983.

Fort Rucker, AL. Home of U.S. Army Aviation, http://www.rucker.army.mil. Accessed April 2002.

Francillon, Rene J. *Vietnam: The War in the Air.* New York: Arch Cape Press, 1987.

Fredriksen, John C. *Warbirds: An Illustrated Guide to U.S. Military Aircraft, 1915–2000.* Santa Barbara, CA: ABC-CLIO, 1999.

Galvin, John R. *Air Assault: The Development of Airmobile Warfare.* New York: Hawthorn Books, 1969.

Gregory, Barry. *Vietnam Helicopter Handbook.* Wellingborough, England: Patrick Stephens Limited, 1988.

Gunston, Bill, and Mike Spick. *Modern Fighting Helicopters.* London: Crescent Books, 1996.

Gurney, Gene. *Vietnam: The War in the Air.* New York: Crown Publishers, 1985.

Harding, Stephen. *U.S. Army Aircraft since 1947.* Stillwater, MN: Specialty Press, 1990.

Heatley, Michael. *The Illustrated History of Helicopters.* New York: Bison Books, 1985.

"Helicopters of the U.S. Army," http://www.geocities.com/capecanaveral/hangar/3393/army.html. Accessed August 2002.

Helicopter World, http://www.helicopter.virtualave.net. Accessed April 2002.

Helicopter's History Site, http://www.helis.com. Accessed June 2002.

Higham, Robin, John T. Greenwood, and Von Hardesty. *Russian Aviation and Air Power in the Twentieth Century.* London: Frank Cass, 1998.

Hirschberg, Michael, and David K. Daley. *U.S. and Russian Helicopter Development in the 20th Century.* Np: American Helicopter Society, International, 2000.

Howze, Hamilton H. *A Cavalryman's Story: Memoirs of a Twentieth-Century Army General*. Washington, DC: Smithsonian Institution Press, 1996.

Hunt, William E. *Helicopter: Pioneering with Igor Sikorsky*. Shrewsbury, England: Airlife Publishing, 1998.

Igor Sikorsky Historical Archives, http://www.iconn.net/igor/indexlnk.html. Accessed October 2002.

International Helicopters, http://www.globalsecurity.org. Accessed August 2002.

Johnson, Lawrence H., III. *Winged Sabers: The Air Cavalry in Vietnam, 1965–1973*. Harrisburg, PA: Stackpole Books, 1990.

Kelly, Orr. *From a Dark Sky: The Story of the U.S. Air Force Special Operations*. Novato, CA: Presidio Press, 1996.

Keogan, Joseph. *The Igor I. Sikorsky Aircraft Legacy: The Chronology of Fixed-Winged and Rotary-Wing Aircraft of Igor I. Sikorsky and the Sikorsky Aircraft Company*. Stratford, CT: Igor I. Sikorsky Historical Archives, 2003.

Kuznetsov, G. I. *Kamov OKB 50 Years, 1948–1998*. Edinburgh, Scotland: Polygon Publishing House, Birlinn Publishing, 1999.

Landis, Tony, and Dennis Jenkins. *Lockheed AH-56A Cheyenne*. North Branch, MN: Specialty Press, 2000.

Lightbody, Andy, and Joe Poyer. *The Illustrated History of Helicopters*. Lincolnwood, IL: Publications International, 1990.

Marshall, Chris. *The Defenders*. London: Aerospace Publishing, 1988.

McGuire, Francis G. *Helicopters 1948–1998: A Contemporary History*. Alexandria, VA: Helicopter Association International, 1998.

Mesko, Jim. *Airmobile: The Helicopter War in Vietnam*. Carrollton, TX: Squadron Signal Publications, 1984.

Novosel, Michael J. *Dustoff: The Memoirs of an Army Aviator*. Novato, CA: Presidio Press, 1999.

Oren, Michael B. *Six Days of War: June 1967 and the Making of the Modern Middle East*. New York: Ballantine Books, 2003.

Palmer, Norman, and Floyd D. Kennedy, Jr. *Military Helicopters of the World*. Annapolis: Naval Institute Press, 1984.

Pearcy, Arthur. *U.S. Coast Guard Aircraft since 1916*. Shrewsbury, England: Airlife Publications, 1991.

Pember, Harry. *Seventy-five Years of Aviation Firsts*. Stratford, CT: Sikorsky Historical Archives, 1998.

Royal Air Force Official Site, http://www.raf.mod.uk. Accessed May 2001.

Royal Navy Official Site, http://www.royal-navy.mod.uk. Accessed May 2002.

Russian Aviation Museum, http://www.2.ctrl-c.liu.se/misc/ram. Accessed November 2002.

Simpson, R. W. *Airlife's Helicopters and Rotorcraft: A Directory of World Manufacturers and Their Aircraft*. Shrewsbury, England: Airlife Publishing Ltd., 1998.

Soviet/Russian Helicopters, http://www.royfc.com/links/acft_coll. Accessed November 2002.

Spenser, Jay P. *Vertical Challenge: The Hiller Aircraft Story.* Seattle: University of Washington Press, 1992.

Stanton, Shelby L. *Anatomy of a Division: The First Cav in Vietnam.* Novato, CA: Presidio Press, 1987.

Swanborough, Gordon, and Peter M. Bowers. *United States Navy Aircraft since 1911.* London: Putnam, 1990.

Taylor, Michael J. H., ed. *Jane's Encyclopedia of Aviation.* New York: Portland House, 1989.

Tilford, Earl H., Jr. *The USAF Search and Rescue in Southeast Asia.* Washington, DC: Office of Air Force History, 1992.

U.S. Air Force Historical Research Agency, http://www.au.af.mil/au/afhra/. Accessed December 2002.

U.S. Air Force Museum, http://www.asc.wpafb.af.mil/museum. Accessed December 2002.

U.S. Army Aviation and Missile Command (AMCOM), http://www.redstone.army.mil/history/aviation/. Accessed February 2003.

Uttley, Matthew R. *Westland and the British Helicopter Industry, 1945–1960: Licensed Production vs. Indigenous Innovation.* London: Taylor and Francis, 2001.

Weinert, Richard P., Jr., and Susan Canedy, eds. *A History of Army Aviation: 1950–1962.* Ft. Monroe, VA: Office of the Command Historian, U.S. Army Training and Doctrine Command, 1991.

Williams, LTG Robert R. U.S. Army, retired. Oral history interview. 2003.

Young, Ralph B. *Army Aviation in Vietnam, 1963–1966: An Illustrated History of Unit Insignia Aircraft Camouflage and Markings.* 2 vols. New Jersey: Huey Company, 2000.

Young, Warren R., et al. *The Epic of Flight: The Helicopters.* Alexandria, VA: Time-Life Books, 1982.

CHAPTER 5

Technological Evolution, 1976–1990

NATO COMBAT SUPPORT HELICOPTERS

After the helicopter proved its viability in the Vietnam War and in the Third and Fourth Arab-Israeli wars, military leaders around the world realized that rotorcraft were tremendous combat multipliers. Advanced technology in composite rotor systems, avionics, weapons systems, and more powerful engines guaranteed new helicopters to be even more efficient, and deadly, than their predecessors. In NATO, the Warsaw Pact, and the so-called unaligned nations, generals and admirals added helicopters to their weapons inventories.

Notwithstanding the certification of the civil version of the Hughes Model 369 on August 24, 1966, and its commercial success, military variants other than the OH-6 did not appear until the mid-1970s. Hughes changed the designation to the Model 500 and continued to improve the aircraft, especially with more powerful Allison engines. In 1976, Hughes introduced the 500D with a 420-horsepower Allison 250-C20B engine driving a new five-bladed main and four-bladed tailrotor, installed on a "T"-tail. In January 1982 the 500E followed with a slightly modified pointed nose. The Model 500MD Defender offered international armed forces a small, heavily armed scout helicopter. Modified with armor plating, an optional telescopic roof-mounted or mast-mounted sight, and external hardpoints for 2.75-inch Folding Fin Aerial Rockets (FFARs), 7.62-mm miniguns, and TOW missiles, the Defender—equipped with the

Allison 250-C20B 420-horsepower engine with IR suppression system—was purchased by several countries. The IDF utilized several export versions of the 500MD to bolster its antiarmor strike force in wars with Israel's Arab neighbors. The Philippines, Colombia, and Mexico all bought versions of the Defender. Korean Air built versions of the 500D, as did Kawasaki for the Japanese Self-Defense Force, and BredaNardi delivered several license-built 500s to the Italian paramilitary Guardia di Finanza and two to the Maltese Armed Forces.

Domestically, the Hughes helicopters returned to prominence in the U.S. military. In the early 1980s, because of the catastrophic failure to rescue the U.S. hostages in Iran, the U.S. Army established the 160th Aviation Regiment to support an independent special operations command within the U.S. military. The Pentagon selected the Army to establish the first aviation special operations unit because of the overall expertise of Army aviators with night vision goggles (NVG). The 160th "Nightstalker" regiment chose, as one of its first aircraft, improved versions of the Hughes OH-6. The OH-6B and OH-6C, or MH-6/C, included more powerful Allison 250-C20 turboshafts, protected by IR suppression kits, and hardpoints for 7.62-mm miniguns and 2.75-inch FFAR rocket pods. The instrumentation included LORAN navigation systems installed in an NVG-compatible cockpit.

Throughout the 1980s, Special Operations Command (SOCOM) continued to upgrade its operational capabilities with aircraft from several manufacturers, including variants of the Hughes, then the McDonnell-Douglas (which bought out Hughes Helicopters in January 1984) 500D and 500E. Referred to as the OH-8, or MH-6D/E, these aircraft had Allison engines of up to 600 horsepower and were employed as troop carriers for covert operations. To provide fire support for these clandestine missions, SOCOM purchased several AH-6F helicopters with 650-horsepower Allison 250-C30 engines, which allowed the machines to carry a variety of armament, including Stinger AAMs, TOW ATGMs, or a .50-caliber machine gun. At least one model reportedly carried a 20-mm cannon.

As the United States withdrew from South Vietnam, the U.S. Army solicited bids to replace the venerable Huey with a larger and faster machine that could carry more troops and supplies to the battlefield. In 1972 Army planners initiated the Utility Tactical Transport Aircraft System (UTTAS) program to develop a utility helicopter capable of transporting an entire infantry squad and its equipment in the most rigorous of environments. The Army selected the Boeing-Vertol YUH-61, the Bell Helicopter Textron Model 214, and the Sikorsky S-70 for evaluation, the winner earning the Army contract for the new utility helicopter.

Bell engineers basically updated and enlarged the Huey for the UTTAS competition. In 1970, Bell had introduced a single proto-type of the Model 214 Huey Plus, powered by a Lycoming T53-L-13 1,400-horsepower turboshaft. The 214 featured a new "Nodomatic" rotor system with a wider chord and no stabilizer bar, and a trans-mission capable of handling the engine's full power. The machine also included structural modifications that allowed the aircraft to lift greater loads. Bell entered a production Model 214A into the UTTAS competition. The 214A resembled the prototype but with a more streamlined nose and powered by a Lycoming LTC4B-8D 2,930-horsepower turboshaft.

The Army selected the Boeing and Sikorsky models for the UTTAS fly-off, but in December 1972, Bell demonstrated the 214A to Iran, and the shah's imperial government immediately placed a large order for its Army, which named the helicopter Isfahan, for the town near the Iranian helicopter training base. The Iranian Air Force followed with an order for thirty-nine 214Cs outfitted for the SAR role. Bell then began discussions with Iran for a 214ST (stretch twin), with two T700 2,250-horsepower engines and other improvements, but in 1979, the year in which three prototypes of the new aircraft flew, Is-lamic insurrectionists overthrew the shah. Bell produced more than 500 military versions of the 214, with 332 sold to Iran, many of which served in the 1980–1988 war with Iraq. In 1976, Bell offered a civilian 214B, but with only a few sales to Dubai, Ecuador, Oman, and the Philippines, production ended in 1981.

Bell decided to revamp its 214 twin to the 214ST Super Trans-port and market it commercially. The first deliveries of the 214ST began in 1982. More streamlined than previous Hueys, the 214ST fuselage was 96 inches longer than the 214 and seated up to eight-een passengers. Two GE CT7-2A 2,350-horsepower turbines turned a 52-foot composite, semirigid rotor system with titanium leading edges and stainless steel end caps. Bell installed a five-bladed rotor system on a few later models. With its large cabin, capable of carry-ing an 8,000-pound payload at 140 knots to a range of 430 nautical miles, the 214ST quickly became popular with commercial opera-tors, especially those involved in offshore oil and gas exploration. Several countries bought military variants of the helicopter, includ-ing Iraq's purchase of 45; there were also a few sales to Brunei, Oman, Peru, Venezuela, and Thailand.

Boeing-Vertol also entered the UTTAS competition with the YUH-61A. A major departure for the company, the YUH-61A was a medium lift helicopter of conventional design that offered three times the payload capability of the Huey. Incorporating a fourteen-man cabin and tricycle retractable landing gear, two General Electric T700 1,500-horsepower turboshafts drove the YUH-61's single four-bladed main rotor and the conventional antitorque rotor. The company also made plans to produce the Model 179, a twenty-five-passenger civilian version of the helicopter. When the Sikorsky design won the UTTAS competition, however, Boeing executives concluded that neither version was viable and discontinued both projects.

Although the U.S. Army planned to replace its UH-1 fleet, the Huey retained its popularity in foreign service. Bell added improved avionics, IR suppressors to engine exhausts, and night vision goggle–compatible cockpits to modernize the Huey. The company sold small quantities of the helicopter to Australia, Canada, Norway, and New Zealand. The Canadians made use of the CUH-1H, later CH-118, for pilot training. Turkey bought at least fifty-five U.S.-produced UH-1Hs, although most foreign-service Hueys were built under license. In the Federal Republic of Germany, Dornier produced 352 UH-1Ds, and in Taiwan, AIDC manufactured 118 UH-1Hs for the Nationalist Chinese Army.

Throughout the last thirty years of the twentieth century Agusta remained the most important source for license-built single- and twin-engined versions of the Huey. Agusta adopted a policy of providing almost any auxiliary equipment a customer desired, such as rotor brakes, additional fuel tanks, improved particle separator to keep sand and dust out of the engines, floats, skis, and a number of armament systems. The company also offered the AB-204 with either a Rolls-Royce Bristol Gnome H.1200 1,200-horsepower or a GE T58-GE-2 1,325-horsepower turboshaft, and the AB-205 with a twin-pack of those engines. As a result of this policy Agusta supplied versions of the Model 204 and 205 to Austria, Greece, Iran, Italy, Morocco, Oman, Saudi Arabia, Singapore, Spain, Sweden, Tanzania, Tunisia, Turkey, Yemen, Zambia, and Zimbabwe. The company also manufactured a specialized ASW variant, designated the AB-204AS. Agusta's engineers equipped this Huey with a search radar, dunking sonar, homing torpedoes, and other maritime weapons systems. Italy, Spain, and Turkey all bought several aircraft for their navies.

In the early 1970s, Agusta obtained a license for the Bell Model 212 and sold numerous transport/utility and SAR versions of the machine to Austria, Dubai, Italy, Lebanon, Morocco, Saudi Arabia, Somalia, Spain, Sudan, Yemen, and Zambia.

In 1973, Agusta introduced the AB-212AS, with twin PT6-6 1,875-horsepower engines. The helicopter resembled a standard Model 212 except for a prominent drum-shaped radome above the cockpit, housing a Ferranti SeaSpray radar. A Bendix ASQ-18 dipping sonar hung from a winch on the aircraft's port side. Two technicians operated the ASW equipment, and a flight direction and navigation system was integrated with the search and fire control systems to pinpoint targets. Two Mk 44/46 or Moto Fides A-244/S homing torpedoes could be fitted to the AB-212AS in the ASW role. For surface attack the aircraft carried two Italian Marte Mk.2, BEA Sea Skua, or Aerospatiale AS-12 light antiship missiles. The AB-212AS also had the capability to carry a data link that provided guidance updates to the Italian Otomat antiship missile. Greece, Iran, Italy, Spain, Turkey, and Venezuela bought copies of the AB-212AS.

In 1982, Agusta introduced a prototype of a multirole military version of Bell's Model 412, the "4," indicating a four-bladed main rotor system, designated the AB-412 Grifone (Griffon). In its various configurations the helicopter might function as a gunship, medevac, troop transport, or SAR aircraft. The Grifone contained armored, crash-resistant seats for the pilot and copilot, high-impact landing gear, self-sealing fuel tanks, and permanent armament stations. The Grifone carried a versatile array of weapons, which might include two .50-caliber heavy machine gun pods, or two 25-mm Oerlikon cannon, along with two 19-round 81-mm SNORA rockets, and up to eight TOW missiles. Stinger AAMs or four BAE Sea Skua antiship missiles might be among the weapons mix, depending on the mission. The air ambulance and utility versions accommodated six stretchers and two medical attendants, or up to fifteen combat-loaded troops. Dubai, Finland, Italy, Lesotho, Uganda, and Venezuela bought various versions of the Grifone.

With continual improvements, the Bell OH-58/206 continued to be one of the world's most popular helicopters during the 1970s and 1980s. In 1976, Austria bought a dozen OH-58Bs, a basic "A" model with modifications to meet Austrian requirements. In the early 1980s, seeking to improve its light utility and scout helicopter capabilities, the U.S. Army began the Army Helicopter Improvement Program (AHIP). In the interim the Army contracted Bell to upgrade almost 500 OH-58As to the OH-58C configuration. Bell installed a more powerful Allison 250C20 420-horsepower engine, an improved transmission, infrared suppression on the engine exhausts, and a modified rounded windscreen to a less reflective flat-plate design. In the 1990s the Army equipped about 100 OH-58Cs to carry the Stinger AAM and added a lightweight airborne GPS to 291 of the aircraft. A twenty-first-century reorganization of U.S. Army Aviation relegated all OH-58s to Initial Entry Rotary Wing (IERW) flight training at Fort Rucker, Alabama.

In September 1981 the Army chose Bell's Model 406 as the winner of the AHIP competition and awarded the company a contract to produce the OH-58D Kiowa. The AHIP featured an innovative four-bladed "soft-in-plane" rotor system that dampened lead and lag of the individual blades but offered more maneuverability than any previous Bell product. An Allison 250-C30R 650-horsepower engine and further improved transmission offered more power and airspeed to the helicopter, which was constructed around a rebuilt OH-58A airframe. The rotor measured 35 feet, 4 inches in diameter; overall length was 40 feet, 11.75 inches. The aircraft attained airspeeds above 140 knots, but the Army restricted maximum speed to 120 knots, which gave the aircraft a range of 300 nautical miles.

A spherical Mast Mounted Sight (MMS) attached above the rotor head distinguished the new helicopter from any previous model. The sight, developed and produced by a team of engineers from Northrop and McDonnell Douglas, contained a high-resolution television camera and thermal imaging system for long-range, day/night target detection, even though targets might be obscured by inclement weather or battlefield conditions. A laser rangefinder/designator, linked to an advanced navigation system, provided precise target location/guidance of HELLFIRE missiles and Copperhead artillery rounds, or handoff to an AH-1 Cobra for TOW missile engagements. To protect the helicopter from enemy SAMs, the Army installed the ALQ-144 IR jammer and the improved APR-39 Radar Warning Device.

The MMS allowed the Kiowa crew to view the battlefield without unmasking (exposing) the helicopter above tree lines or terrain, making the aircraft almost invisible to an enemy.

The OH-58D cabin configuration eliminated the backseat and filled the area with an array of computers, making the new Kiowa one of the most sophisticated in the Army inventory. With the first all-glass cockpit on a U.S. Army helicopter, the primary multifunction displays (MFD) provided the crew with vertical and horizontal situation information, communications control, and MMS video. The OH-58D provided the crew with the capability to preplan their missions and data-store all navigation waypoints and mission frequencies in onboard computers. Switches on the cyclic and collective allowed the crew to select weapons, radio frequencies, and flight/navigation displays without releasing the flight controls. With folding blades and a quick-removal system for the MMS, the Kiowa could be rapidly onloaded aboard a transport aircraft, flown anywhere in the world, and be ready to fly within minutes of off-loading. The distinctive ball above the rotors, with its two round aperture, earned the helicopter the nickname of "ET" ("extraterrestrial"), after a popular movie of the late 1970s.

In 1987 conditions in the Persian Gulf created a requirement for a small, armed helicopter to interdict terrorist gunboats attacking maritime transportation, especially tankers, at night. In less than three months Bell Helicopter, coordinating with a U.S. military team, produced an armed version of the OH58D, the Kiowa Warrior. The DoD deployed fifteen of these aircraft to the Persian Gulf, stationing the helicopters, designated Task Force 118 "Prime Chance," aboard Navy ships with the mission of protecting the critical oil supply routes. The Kiowa Warrior crews devastated the speedy gunboats, and, after only two engagements, the boat crews refused to venture out to attack shipping or mine the sea lanes.

As a derivative of the OH-58D, Bell introduced the Model 406CS Combat Scout in the mid-1980s. Intended for sales to foreign governments, the Combat Scout came equipped with either a roof-mounted sight or the MMS installed on the OH-58D. Armament included combinations of 7.62-mm or .50-caliber machine guns, 2.75-inch FFAR pods, and TOW or Hellfire missiles.

In 1986, Bell transferred production of the Model 206 to the new Textron Canada factory at Montreal-Mirabel. The facility also produced the Model 206L Long Ranger, which has a lengthened fuselage to provide club seating for three additional passengers in the executive model. Additionally, the company modified the 206L into the Model 407 with a four-bladed main rotor system, improved transmission, advanced avionics, and a ring-shroud to protect the tailrotor. Bell's Model 427, a 407 with a new transmission system, twin Pratt & Whitney PW-206D engines, and an enlarged cabin, appeared in the mid-1990s. Tridair Corporation offered a twin-engine conversion for the Jet Rangers with the "Gemini ST" configuration utilizing a Soloy Dual engine installation. Bell also marketed the Model 206LT, an equivalent twin-engined conversion.

In the mid-1970s, Bell Helicopter Textron introduced a completely new executive helicopter designed for the commercial market. The Model 222, like the Jet Ranger, had a separate cockpit and cabin with seating for eight to ten passengers. Two Lycoming LTS101-650C-3 592-horsepower turboshafts turned the two-bladed main and tailrotors. Uniquely, the 222 featured a tricycle-wheeled landing gear that retracted into sponsons along the lower stream-lined fuselage. Bell upgraded the 222B with a larger rotor system and two LTS101-750C-1 685-horsepower engines. Generally mar-keted as a business aircraft, the 222 found acceptance among civil governments and medical services. Several police forces, including the London Metropolitan Police, adopted the 222. The company also delivered the Model 222UT utility version, equipped with skid gear, to several foreign militaries. In the early 1980s, ironically, the helicopter designed for business executives earned fame as a mili-tary helicopter in the television series *Airwolf,* in which the helicop-ter performed impossible maneuvers and demonstrated futuristic weaponry. Bell upgraded the Model 222 to the Model 230, which was followed by the higher powered Model 430 equipped with a re-designed four-bladed main rotor system. The Model 430, along with all Jet Rangers, continued to be manufactured at Mirabel.

Sikorsky entered its S-70 into the U.S. Army UTTAS competition and eventually won a contract to produce the Army's new combat as-sault helicopter. In the midst of developing the new aircraft, on May 1, 1974, Sikorsky Aircraft became a division of the newly estab-lished United Technologies Corporation. Designed to carry a crew of three and a fully equipped eleven-man U.S. infantry squad, the UH-60A Blackhawk emerged the winner of a rigorous evaluation and intensive seven-month fly-off in December 1976. The Army in-tended to order more than 1,100 UH-60s as troop and cargo trans-port, medevac (with four litters), reconnaissance, and command and control aircraft.

On October 17, 1974, the first Sikorsky prototype took flight, powered by two General Electric T700-GE-700 1,620-horsepower turboshafts, turning a composite titanium and fiberglass 53-foot, 8-inch four-bladed main rotor system, and a four-bladed tailrotor. Sikorsky engineers equipped the UH-60 with the latest in technol-ogy, including IR suppression kits for the engines; Stability Control Augmentation System (SCAS); transmission and gearboxes with thirty-minute dry-run capabilities; rotors and driveshafts designed to sustain multiple hits by 23-mm cannon fire and remain operational; armored, crash-resistant crew seating and energy-absorbing wheeled landing gear; and deicing capability to operate in moderate icing conditions. In the fly-off prototypes reached 195 knots, lifted off at

21,000 pounds gross weights, hauled a 7,000-pound slingload, and demonstrated a service ceiling of 18,400 feet.

During the UTTAS operational test program, vibrations forced an Army crew to execute a night landing in the thick forests of Fort Campbell, Kentucky. The Blackhawk's main rotor blades chopped down 12-inch pine trees, and the aircraft landed with no injuries to the crew or infantrymen aboard. After judging that the UH-60 had sustained only superficial damage, an Army/Sikorsky support team installed a new set of rotor blades and enlarged the LZ with a chain saw, allowing the Blackhawk to complete the evaluation. The ruggedness of the UH-60 duly impressed Army evaluators, affecting their decision to select the UH-60 as the winner of the UTTAS competition. The Army accepted delivery of its first Blackhawk in August 1978 (Pember 1998, 60).

In the early 1980s, Sikorsky also modified several UH-60s to meet other, diverse Army requirements. The company modified several UH-60As for special operations with (Forward Looking Infra Red) FLIR, door-mounted 7.62-mm miniguns and auxiliary fuel tanks. The EH-60E Quick Fix II, Electronic Countermeasures (ECM), replaced the UH-1 Quick Fix system fielded in the 1970s. In addition to an advanced navigation system and avionics, the new variant carried a unique external dipole antenna integrated into a computerized Radio Direction Finding (RDF), radio intercept, and communications jamming system. The EH-60B Stand-Off Target Acquisition System (SOTAS) helicopter carried an underfuselage Side-Looking Aerial Radar (SLAR) antenna, designed to detect movement of enemy forces, especially vehicles and low-flying aircraft. The SOTAS flew behind friendly lines, and the radar's Moving Target Indicator (MTI) detected enemy movements and relayed the intelligence to a ground station via a digital datalink. A "long-stroke" landing gear permitted the SLAR antenna to rotate 360 degrees beneath the aircraft. A specially equipped medevac, the UH-60Q, provided enhanced medical care for six litter patients, with an onboard oxygen generating system, electronic heart monitors, and a medical suction system. Saudi Arabia bought eight versions of the Desert Hawk, an UH-60Q with specialized avionics and air conditioning.

Blackhawks saw significant combat service in both Granada and Panama in the 1980s. Both regular Army and Special Operations units utilized the UH-60 and were impressed with the rugged Sikorsky aircraft. In Operation Urgent Fury, the 1983 U.S. invasion of Grenada, several Blackhawks suffered multiple hits by 12.7-mm and

23-mm cannon fire but returned their crews to safety. A few Blackhawk pilots landed their bullet-riddled, smoking aircraft on Navy
carriers. In one instance the engine controls were shot away, and the
Navy firefighters had to shut the Blackhawk engines down by running a firehose into the engine inlets before they could shove the
hulk overboard.

In February 1978 the U.S. Navy ordered the SH-60B Seahawk,
Light Airborne Multi Purpose System (LAMPS), helicopter with GE
T700-GE401 turboshafts, folding main rotor blades and tailboom,
and a longer-life control system. Sikorsky equipped the Seahawk
with typical ASW equipment of a dipping sonar, twenty-five
sonobuoys, and two to three Mk 46/50 acoustic torpedoes. The
Navy later updated the Seahawk with an IBM LAMPS III ASW
suite that included the antiship Penguin missile, T700-GE-401C
1,900-horsepower engines, extra fuel tanks, and added emergency
flotation equipment. In the early 1980s the Seahawk began ASW
operations from Navy frigates, destroyers, and cruisers. Additional
missions for the SH-60 B included SAR and vertical replenishment.
The Navy subsequently ordered versions of the S-70 for plane guard
and special operation missions. Designated HH-60H, SH-60F, and
SH-60FST, the latter intended for special warfare, delivery/extrac-
tion of SEAL teams, and combat rescue operations, they were
equipped with radar warning and ECM equipment, .50-caliber
doorguns, and mounting points for Hellfire missiles. The Navy
planned on further orders for a multirole helicopter designated SH-
60R. The USMC ordered a specialized VIP version of the SH-60B
for its Presidential Flight Detachment with weather radar, secure
communications equipment, and a luxurious executive interior.

The USCG procured the 180-knot HH-60J JayHawk Medium
Range Rescue (MRR) helicopter for SAR, law enforcement, and environmental control operations. The JayHawk came equipped much
like the SH-60, but with a nose-mounted search radar and GPS navigation system. Although it was not able to land on water, the Jay-
Hawk could range out to 300 nautical miles, hover for forty-five
minutes, and return to base without becoming fuel critical.

On March 13, 1977, Sikorsky introduced the S-76 Spirit, so designated to coincide with the U.S. bicentennial. A high-performance
twelve-passenger executive aircraft, the Spirit was the first helicopter the company developed solely for the civilian market. The Spirit
featured extensive use of composites, with an integral fuselage, tailboom, tailfin, and retractable tricycle landing gear. Powered by two
Allison 250-C30 650-horsepower turboshafts, which drove the four-

bladed composite main rotor, the 10,000-pound Spirit cruised at 130 knots and had a range of 385 miles. Sikorsky followed the initial aircraft with models equipped with the more powerful Turbomeca Arriel 700- and 850-horsepower and Pratt & Whitney PT6B-36B 980-horsepower engines, along with improved transmissions that increased gross weight to 11,700 pounds and cruise speed to 155 knots.

In 1985, Sikorsky followed with the H-76, later AH-76, Eagle with an NVG-compatible cockpit, shrouded tailrotor, and armament hardpoints. Typical Eagle armament included 7.62-mm miniguns, 2.75-inch FFAR rocket pods, and TOW missiles, but few countries expressed interest in the military version. The civilian S-76, however, enjoyed considerable success, especially with offshore oil operators and EMS services requiring a larger helicopter. Certified as a single-pilot IFR aircraft, the Spirit could lift a 5,000-pound slingload. Sikorsky produced more than 1,100 of all models of the S-76.

In the early 1970s, Aerospatiale designed a light utility helicopter to compete on the international civilian market with the Bell 206 Jet Ranger. The AS 350 Ecureuil (Squirrel) incorporated minimal, simple parts to reduce production costs and to increase maintainability. To that end, Aerospatiale used the three-bladed Starflex rotor head and a fuselage composed of light composites and alloys. The Ecureuil prototype made its first flight on June 27, 1974, and received its French certificate in October 1977. The company marketed the AS 350 in North America as the A-Star, equipped with a Lycoming LTS101 engine, and elsewhere with the French-built Turbomeca Arriel 1A turboshaft. Capitalizing on the success of the AS 350, Aerospatiale introduced the twin-engined AS 355 Twin-Star in the United States, or the Twin Squirrel in Europe. The machine cruised at 155 knots for 362 nautical miles and could lift a 2,000-pound slingload.

Aerospatiale developed an armed variant of the AS 350 equipped with machine guns and FFARs, or four TOW missiles, or a single 20-mm cannon, but with only two hardpoints, ordinance load was limited. Denmark and Singapore ordered antitank versions of the aircraft, redesignated AS 550/555 Fennec in 1990, and Australia ordered unarmed versions for the Royal Australian Air Force (RAAF) and Royal Australian Navy (RAN) to replace several UH-1B trainers. During Desert Storm three RAN Fennecs, stationed aboard frigates and armed with a 7.62-mm doorgun, conducted reconnaissance missions in the Persian Gulf. Both types were produced in Brazil, and China's CHAIC Z-11 appeared to be a direct copy of the French machine. At the end of 2003 production and orders for both the single- and twin-engined versions worldwide topped 2,250. In 2003, fifty-one of the helicopters appeared on Australia's aircraft registry, used for such diverse tasks as news reporting, law enforcement, agriculture, SAR, and as support for Australia's Antarctic stations. Proving the international popularity of the AS 350/550/5555, Albania, Argentina, Australia, Botswana, Brazil, Cambodia, China, Denmark, Guinea, Ireland, Jamaica, Mexico, Nepal, Paraguay, Peru, Sierra Leone, Singapore, Thailand, Tunisia, and the United Arab Emirates all operated variants of the helicopter.

First flown in 1965, the Aerospatiale-designed "Puma-type" aircraft remained in production until 1987, but in 1974 the company initiated an improvement program that led to the AS 332 Super Puma. In September 1977 a prototype SA-331, equipped with an improved transmission and two 755-horsepower Turbomeca Makila turboshafts, completed its first test flight. On September 13 of the next year the prototype AS 332 Super Puma appeared with increased engine power and a more aerodynamically efficient fuselage. Aerospatiale engineers installed new composite main and tail-rotors that increased lift efficiency, service life, combat survivability, and ended the problem of metallic blades corroding in maritime operations. The Super Puma featured an extended fuselage capable of hauling troops and cargo, a wider energy-absorbing retractable landing gear, armored seating for passengers and crew, and greater fuel capacity. The AS 332 reached a range of 532 nautical miles at maximum speeds of 141 knots. From 1978 until 1987 the company introduced five military and civilian variants of the 332, two with extended fuselages. In 1980 the 332 replaced the standard AS 330 Puma as the primary utility helicopter produced by Aerospatiale. The company manufactured 670 of the type, and the aircraft remained in limited production in Rumania into the twenty-first century.

Two fuel-efficient Turbomeca Makila 1,877-horsepower turboshafts turned the 51-foot, 2.2-inch four-bladed main rotor mounted atop the elongated rectangular fuselage, accommodating up to twenty-four troops. A longer tailboom and vertical fin supported a four-bladed composite tailrotor mounted on the starboard side and a horizontal stabilizer on the port. Depending on the version, in addition to more troops and twelve litters, or SAR equipment, the Super Puma could carry a 10,000-pound slingload, or an assortment of weapons for land or sea warfare. Weapons included gun pods for 7.62/.50-caliber machine guns, or 20-mm cannon and 70-mm

FFARs. The ASW version carried a search radar in an extended nose, a dipping sonar, and homing torpedoes. For antiship operations the aircraft carried two AM 39 or MM 40 Exocet, BEA Sea Skua, or Harpoon missiles. Transport versions were armed with pintle-mounted 7.62-mm or .50-caliber doorguns. Bristow Helicopters Ltd., acquiring more than thirty aircraft for offshore operations in the North Sea and IPTN (Industri Pesawat Terbong Nasantara) in Indonesia, which also built the SA-330 under license, manufactured several Super Pumas for domestic civilian and military use. In all, thirty-eight countries bought variants of the AS 332-532, including Algeria, Argentina, Belgium, Brazil, Chad, Chile, China, Djibouti, Ecuador, France, Gabon, Germany, Iceland, Indonesia, Iraq, Ivory Coast, Jordan, Kenya, Kuwait, Lebanon, Libya, Malawi, Mexico, Morocco, Nepal, Nigeria, Oman, Pakistan, Panama, Philippines, Portugal, Senegal, Singapore, South Africa, Spain, Sudan, Switzerland, Togo, United Arab Emirates, the United Kingdom, and Zaire.

In the early 1980s the U.S. Coast Guard began to search for a helicopter to replace its aging Sikorsky S-62/HH-52A fleet. To meet that requirement Aerospatiale contracted to produce the SA-366, a modified version of the SA-365N Dauphin II. The helicopter, designated the HH-65A Dolphin by the USCG, measured 44 feet, 5 inches in length, with a rotor diameter of 39 feet, 2 inches. Two Arriel turboshafts powered the prototype. Aerospatiale designed the helicopter's retractable tricycle landing gear to land aboard USCG cutters and icebreakers operating in rough seas, but the aircraft are generally shore-based and effect rescues and searches out to 150 nautical miles offshore. Designed for short-range rescue missions, the advanced computerized communication, navigation, and flight management suite allowed the pilot to program the system to fly the helicopter from a known point and arrive at a 50-foot hover above a selected destination. Additionally, the advanced flight-direction system flew preselected search patterns allowing the crew to concentrate on spotting the object of the search. Capable of carrying its crew of four at 165 knots, the HH-65A had a range of 400 nautical miles. In addition to SAR the USCG used the Dolphin in law enforcement, especially drug interdiction, and vertical replenishment of ships and isolated communities in polar regions. The USCG took delivery of the first of ninety-six HH-65As in 1985. The Turbomeca engines proved underpowered, and U.S. laws required a certain percentage of U.S.-made components for a government purchase; the USCG therefore prescribed that the Textron-Lycoming LTS.101-750B-2 700-horsepower turboshaft replace the French engines.

On April 2, 1982, Argentinean forces invaded the Falkland Islands, touching off the Falklands War, or the *Guerro Pour Los Malvinas,* which lasted until June 20. Both countries possessed many of the same type of helicopters, but, despite the loss of most of their helicopters when an Exocet missile slammed into the container ship *Atlantic Conveyor,* the British helicopters and crews proved much more effective than the Argentineans. Operating in immoderate weather conditions, the UK machines accomplished extraordinary rescues and inflicted heavy losses on the enemy. When two Wessex HU5s crashed on Fortuna Glacier, another Wessex crew, contending with 90-mph winds and blinding snow, rescued both downed crews. During the short war, Wessex and Sea King helicopters plucked several downed aircrews from the icy waters off the Falklands, in both plane guard and SAR roles. In their baptism of fire, 20 naval Lynxes, armed with Mk 44, Mk 66, or Stingray torpedoes, or four BEA Sea Skua ASMs, scored 100 percent accuracy in the antisurface role. The Lynx HAS 2 was faster and more agile than previous British ASW helicopters and carried an updated Sea Spray search/targeting radar to locate enemy shipping and the 600-mph Sea Skua, designed to attack vessels moving at up to 50 knots. Army Lynxes, and the Chinook saved from the fire-ravaged *Atlantic Conveyor,* performed more than credible service in transporting troops and supplies throughout the campaign.

Both the British and Argentinean military lost helicopters during the war, in combat and operational accidents. Ground fire shot down British Sea Kings and Argentinean Pumas, and both sides lost helicopters when ships on which they were based sank as a result of naval combat. Argentine Pucara ground attack aircraft shot down a couple of Gazelles, killing the crews, and a friendly fire incident, when HMS *Cardiff* mistakenly shot down another Gazelle with a Sea Dart, cost the United Kingdom another aircraft and crew. British Sea Harriers shot down at least three Argentinean Pumas and destroyed two others plus an Agusta 109 on the ground. By the end of the conflict British forces had captured nine Bell UH-1H Hueys, two 212s, and several Pumas left on the islands.

Throughout the 1980s, Westland continued to improve the Lynx. An attack version of the British Army Lynx's increased firepower included two 20-mm cannon pods, and a combination of either HOT, Hellfire, or TOW ATGMs, along with 7.62-mm doorguns. More significantly, August 9, 1985, marked the first flight of a specially modified Lynx for the British Experimental Rotor Program (BERP). The new "paddle-shaped" composite rotor blades diverted the tip vortices down and away from the flight path, so that the advancing blade passed through "clean," undisturbed air, thus producing more efficient lift and less drag. The Soviet Mi-24 Hind held the world helicopter absolute speed record of 198.18 knots (367 kph) over a closed course. On August 11, 1986, a "G-Lynx," equipped with a five-bladed BERP system, shattered the record by posting an airspeed of 216.47 knots (400.87 kph). Monitored by cameras and radar, the undisputed record still stands.

Although seeking commercial certification for the Lynx seemed financially imprudent, Westland executives postulated that there was a market for a larger version of the helicopter. Thus they authorized the development of the WG 30. The new rotorcraft combined the dynamic components of the Lynx, including a 42-foot composite rotor system, with two Gem 41-2 1,260-horsepower engines and a redesigned 52-foot fuselage. The craft was designed primarily as a civil helicopter, and its engines sat above the passenger compartment, holding seventeen; there was also a roomy, separate cockpit for two pilots in the forward section. A high-set, twin-finned tailboom supported the tailrotor. The company privately funded the Westland 30's production, counting on orders from British Airways Helicopters, Airspur of Los Angeles, and Omniflight, which operated Pan Am shuttles from downtown New York City to JFK Airport. Despite upgraded GE CT7-2B turboshafts and the five-bladed BERP rotor system, engine problems and a series of accidents in India terminated the Westland 30's service in the United States, as well as other aircraft purchased for international oil and gas support work. The company attempted to develop a military version of the Westland 30, but the very successful Puma negated any interest in a new medium lift helicopter. In 1988 the company discontinued the Westland 30 after producing only forty-one aircraft.

ATTACK HELICOPTERS

During the early 1970s, the U.S. Army conducted a series of force on force tests near Ansbach, Germany. Intended to determine the suitability of air cavalry and attack helicopter units, and the AH-1Q Cobra in particular, to operate in an antitank role in the European environment, these tests pitted helicopter OH-58/AH-1 teams against U.S. armored units familiar with the European terrain. The tests aptly demonstrated that antiarmor helicopter teams, properly trained and employed, achieved high ratios of armored vehicles destroyed for every missile-firing helicopter lost. In several engagements the Cobras scored kill ratios of over 10 to 1. The 3,000-meter and later 3,750-meter range of the TOW allowed AH-1s to score hits in tank kill zones with the tank crews completely unaware of the helicopter's presence. Armed scout and attack helicopters, especially when operating in nap-of-the-earth (NOE) and nighttime environments, clearly demonstrated a viable battlefield survivability as well as being an essential combat multiplier in mid- to high-intensity warfare. With Warsaw Pact tanks outnumbering those of NATO by five to one, the "Ansbach Tests" provided the impetus for modernization of U.S. attack and scout helicopters, as well as reforming doctrinal principles that the Army would take into the next century.

In 1976, as a result of the Ansbach Tests, the Army began a three-step program to upgrade all AH-1G/AH-1Q/AH-1R Cobras to the advanced modernized version, the AH-1S. All but the final new production model of the upgraded Cobras utilized the M73 reflex sight for rocket firing and a Telescopic Sight Unit (TSU) for firing the TOW missile system and turret weapons. The AH-1S actually fell into four categories: the AH-1S Modified AH-1G, the AH-1S Production model, the AH-1S Enhanced Cobra Armament System (ECAS), and the AH-1S Fully Modernized Cobra. The Army soon unofficially changed the designations to AH-1M, AH-1P, AH-1E, and AH-1F, to discriminate between the different Cobra models and their capabilities. None of the four weapon subsystems installed on the four AH-1S variants were interchangeable without considerable modifications to the aircraft, although all the TOW-equipped AH-1s could carry up to eight missiles.

With the installation of the XM65 TOW/Cobra missile subsystem, the AH-1S's primary mission shifted to an antitank role, while retaining the direct aerial fire support, armed escort, and reconnaissance missions. Although it had a strengthened tailboom and improved gearboxes, only the flat-nose mounting of the TSU distinguished the Modified Cobra from the AH-1G, as the modified AH-1S also had the traditional rounded cockpit canopy. Beginning with the AH-1s Production model (AH-1P), the new flat-plate canopy became standard on all Cobras.

In March 1978, after beginning modification of several AH-1Gs and AH-1Qs to the AH-1S configuration, the Army ordered 100 new production Cobras. In addition to the seven-plate flat canopy, Bell installed an improved T-shaped instrument panel arranged to ease the pilot workload of NOE flight. A single T53-L-703 1,800-horsepower turboshaft, and a new transmission and gearboxes, turned improved composite rotor blades. The Production AH-1S carried the XM65 TOW system upgraded with the M128 helmet sight subsystem (HSS) to facilitate firing the new M28A3 armament subsystem, composed of an M134 minigun and M129 grenade launcher. The "Prod" retained the 2.75-inch rocket system.

In September 1978 the U.S. Army initiated step two of the Cobra upgrades by ordering ninety-eight new production AH-1 ECAS Cobras. The new AH-1 retained the M-65 system, but a new 20-mm/30-mm universal turret, with a three-barrel M197 20-mm cannon replaced the M28A3 system. An automatic compensation system stabilized the aircraft for firing the 20-mm off centerline axis. The ECAS also included a new rocket management system (RMS) to select the various 2.75-inch rockets available to the pilot. New tapered, tipped K-747 fiberglass main rotor blades manufactured by the Kaman Corporation provided resistance to 23-mm cannon fire, reduced the noise created by tip vortexes, and increased available lift. The NVG-compatible cockpit included an new Doppler radar navigation system. Several AH-1Es carried the strap-on M130 flare and chaff dispenser.

The final step in modernizing the Cobra fleet began in November 1979 by converting 387 AH-1Gs to AH-1S Modernized Cobras, and buying 143 new Production aircraft. A T53-L-703 1,800-horsepower turboshaft turned the two-bladed K747 main rotor, and extensive use of light alloys and composites lightened the production AH-1F considerably. The FMC retained the instrumentation and Doppler navigation system introduced in the ECAS. In addition to the M65 TOW system, and the M197 20-mm cannon, the Modernized Cobra featured a M26 Fire Control Computer (FCC) linked to a M143 Air Data Subsystem (ADS, mounted on the right side of the canopy), a pilot's M76 Head-Up Display (HUD), M136 pilot and gunner Helmet Sight Subsystem (HSS), laser rangefinder and tracker, and the M147 RMS for the 2.75-inch FFARs carried in either M260 7-tube or M261 19-tube rocket pods designed specifically for the new Hydra 70 multipurpose warheads. In addition to an improved engine IR suppression system and the M130 chaff dispenser, an ALQ-144 infrared jammer sat on the engine fairing just forward of the engine exhaust. Just over fifty AH-1F Cobras received the infrared C-NITE sight upgrade, which allowed the copilot/gunner to acquire and engage targets at night, through adverse weather conditions, or battlefield obscurants. Most of the C-NITE Cobras were deployed to South Korea. Both Army and Marine Cobras participated in combat actions on the island republic of Grenada and in Panama.

In the early 1970s, as a result of its experience in Vietnam, the U.S. Marine Corps identified a need for enhanced attack helicopter capabilities. As a result the USMC ordered an upgraded version of the AH-1J designated the AH-1T. The AH-1T improvements included dynamics from the Bell-produced King Cobra that never reached production. The AH-1T had a longer fuselage and tailboom, new engines, and an upgraded transmission. The "T" model also included the M-65 TOW missile system, providing the USMC with increased firepower for armed escort, direct ground support, and, most important, an antiarmor capability lacked by the AH-1J. In addition to the M-197 20-mm cannon, TOWs, and 2.75-inch FFARs, the AH-1T could also fire the 5-inch Zuni rockets to attack heavier fortifications. On October 25, 1983, in the first significant U.S. military action since Vietnam, AH-1T Cobras covered the Marine amphibious assault on Grenada. Of the nineteen U.S. citizens killed in Operation Urgent Fury, two were Marine aviators, who died when 14.5-mm and 23-mm antiaircraft fire shot down two Marine Cobras over the beaches of Grenada.

On November 16, only two weeks after the end of combat in Grenada, Bell test pilots flew the first in the next generation of Marine Cobras, the AH-1W Super Cobra. The AH-1W provided the Marines with an attack helicopter capable of operating day/night and in adverse weather conditions. Insufficient funding during the late 1970s reduced all U.S. military aircraft inventories, and the Marines sorely needed an upgraded attack helicopter. The USMC contracted for forty-four new AH-1Ws and upgrades for the remaining AH-1Ts to the AH-1W. Two General Electric T700-GE-401 1,690-horsepower turboshafts drove the 48-foot composite rotor system. The Super Cobra possessed a level speed of 170 knots and a range of 317 nautical miles. Armament comprised the 20-mm cannon, either TOW or Hellfire ATGMs, 2.75- or 5-inch rockets, and, significantly for self-defense, AIM-9 Sidewinder AAM; by the mid-1990s it had the new AGM-122 Sidearm Anti Radiation Missile for protection against radar-controlled antiaircraft guns and SAMs.

After the failure of the AH-56 Cheyenne and several Cobra improvements, the U.S. Army still required an attack helicopter capable of unlimited day/night, all-weather antiarmor operations. Thus, the Army invited Request For Proposals (RFP) for an Advanced Attack Helicopter (AAH) program in 1972. The Army selected the Bell

Model 409, YAH-63, and the Hughes Model 77, YAH-64, for a competitive fly-off conducted in 1976. Two General Electric T700-GE-700 1,536-horsepower turboshafts, with IR plume suppressers, powered the YAH-63. Instead of a new rotor system Bell executives chose to provide their machine with a wide-chord, two-bladed semirigid main rotor. The two-bladed stainless steel tailrotor was mounted to a conspicuous vertical fin with a horizontal stabilizer atop the fin. Armament included a General Electric XM188 three-barrel 30-mm cannon mounted in the same universal chin turret used on the AH-1. A stabilized sight mounted under the nose held an FLIR, optics, and laser rangefinder, providing both day and night capabilities. The fire control system included both helmet sights for the tandem-seated pilot and copilot and an autonomous direct targeting system for the copilot/gunner in the front seat. In addition to the cannon, armament included up to sixteen TOW missiles and up to seventy-six 2.75-inch FFARs. The aircraft included armor for the crew, 30-mm ammunition bay, and critical engine components. Bell produced only two prototypes for the fly-off, which the Hughes YAH-64 eventually won.

The Hughes YAH-64, officially designated Apache, also featured two tandem crew stations, with the pilot in the rear. Two General Electric T700-GE-701 1,698-horsepower turboshafts, mounted in nacelles on either side of the fuselage, powered the Apache, which was equipped with a four-bladed fully articulated main and four-bladed tailrotors. The prototype's tailrotor was mounted below a horizontal stabilizer attached high on the ventral fin, but Hughes engineers reversed the installation on production models. Lightweight boron armor, capable of withstanding 23-mm cannon hits, protected the crew and critical components. The Apache sat on a wheeled tricycle landing gear.

The AH-64's computerized fire control system included a Northrop Target Acquisition and Designation Sight (TADS), housing conventional Day Vision Optics (DVO), FLIR, Low-light Level Television (LLTV), a laser rangefinder/designator (LRF/D), and a laser tracker, all integrated with a pilot's night vision sensor (PNVS). The PNVS, operating with a passive IR system, allowed the pilot to fly the aircraft through darkness and inclement weather, and provided a sighting reticle for the cannon. The Honeywell Integrated Helmet and Display Sighting System (IHADSS) provided the crew with navigation and targeting symbology, as well as a "hover box" designed to give the crew a ground reference during conditions of low visibility.

Hughes armed the YAH-64 with the company-designed XM230E1 30-mm chain gun, which experienced ammunition feeding problems in early models. The production Apache carried up to sixteen laser-guided AGM-114A Hellfire antitank missiles, and up to seventy-six Hydra 70 2.75-inch multipurpose FFARs on four articulated wing pylons. To extend the aircraft's range, from one to four 230-gallon auxiliary fuel tanks could be installed on the wing pylons.

Although it first flew on September 30, 1975, almost ten years of development passed before the AH-64 became operational. The Apache went into production in March 1982, with initial deliveries beginning in January 1984, about the same time that McDonnell-Douglas acquired Hughes Helicopters. In 1986 the 6th Cavalry Brigade at Fort Hood, Texas, became the first operational unit with the AH-64A. The 1st Cavalry Division and 101st Airborne Division soon received their quota of the 150-knot attack helicopter.

Flying a helicopter is extremely difficult, especially maintaining the aircraft at a stabilized hover at a fixed point a few feet off the ground. Hovering a helicopter requires a pilot to remain cognizant of his vertical and horizontal situations, while simultaneously manipulating the aircraft controls with both hands and both feet. One fledgling helicopter pilot made an analogy between learning to hover and trying to learn to ride a unicycle on top of a basketball. The Apache's sensors exacerbated the complexity of flying a helicopter, especially at night. An Apache pilot had to mentally transpose his visual perceptions to a point 10 feet forward and 3 feet lower than his head, where the PNVS turret was mounted on the helicopter's nose. The PNVS sensor sent a processed thermal image to the pilot's Integrated Helmet and Display System (IHADS), which included a mini-TV screen positioned just in front of the pilot's right eye. After mastering the transition to flying "the system," AH-64 crews, nonetheless, became a devastating force on the battlefield. With a reduced heat signature and its radar-absorbing paint, the night cloaked the Apache, while its crew flew through the darkness, engaging an enemy that had no idea that the Apache lurked in the gloom. AH-64 crews also participated in SAR operations, using their thermal imaging sensors to locate lost children and flood victims. The Apache is air transportable in USAF C-141, C-5, and C-17 aircraft. In the 1989 Operation Restore Hope in Panama, the AH-64A played a major role at night, overwhelming Panamanian government forces with its firepower.

Although the civil version of the AS 350 earned global respect, the armed version met little success. Aerospatiale continued to pro-

duce the Fennec Antitank Helicopter (ATH), but the machine exhibited several shortcomings. Designed to operate day or night and
in inclement weather, the roof-mounted TOW missile optics restricted the aircraft's antiarmor operations to daylight hours in rea-
sonably good weather. Fog, rain, dust, smoke, or any other battlefield obscurant limited the range of the TOW, normally 3,750
meters. Additionally, the Fennec, with only armored crew seats, was
vulnerable to small arms fire and had no weapons for self-protection. The armed AS 550 attained a maximum airspeed of only 100
knots and a range of 235 nautical miles, and as a troop transport the
Fennec was limited to only four or five passengers. As a result of
these deficiencies Aerospatiale, later Eurocopter, began to develop a
faster, more versatile attack helicopter.

Using the AS 365N as a platform, Aerospatiale developed the
AS365K, later 565 Panther. A fully militarized light utility helicopter, the Panther carried one or two pilots and up to twelve troops, or
3,525 pounds of cargo, or in an armed version, an array of weapons
and targeting systems. On February 29, 1984, powered by two Turbomeca 913-horsepower turboshafts turning a fully composite five-
bladed 39-foot, 2-inch rotor system, the first prototype took flight.
The Panther featured a slim fuselage, capable of hauling twelve
troops or 3,525 pounds of cargo, a fenestron, and retractable landing gear. Reaching a maximum airspeed of 155 knots and a range of
464 nautical miles, the naval version's armament included two AS-
15TT antiship missiles, homing torpedoes, or depth charges. The
French Army armed their Panthers with combinations of 7.62- or
12.7-mm machine guns, 20-mm cannon pods, and the HOT ATGM
system. Later variants also carried either the Mistral or Stinger AAM
for air-to-air self-defense. HAMC in the People's Republic of China
built the SA 365M under license with the 748-horsepower Arriel
1M1 turboshaft, as the Z-9. At least fifteen countries placed orders
for 500 versions of the Dauphin II/Panther helicopter.

In the late 1970s both Messerschmitt-Bölkow-Blohm group
(MBB) and Kawasaki Industries initiated programs to produce a
twin-engined multipurpose helicopter, which resulted in the BK-
117. MBB designers created the BO 107, and Kawasaki produced
the KH7, but the companies reached an agreement to jointly produce a helicopter intended primarily for commercial markets. MBB
provided a 36-foot, 1-inch version of the BO 105 with four-bladed
rigid main- and two-bladed tailrotors, along with the hydraulic and
control systems, while Kawasaki produced the airframe and fuselage
(including the electrical system and interior), transmission, and

landing gear. The design called for two engines mounted atop a fuselage that included rear clamshell doors for easy access to the interior for cargo or patient handling in the EMS version. The 117B2, with two Avco (Textron) Lycoming LTS 101650B1 550-horsepower engines, cruised at 135 knots and hauled up to 2,600 pounds of cargo, either internally or on a cargo hook. The passenger version seated up to eleven, but usually only seven in comfortable executive seats. The EMS version carried two stretchers and two medical attendants with all their equipment. Standard range was 290 nautical miles, or 380 with internal long-range fuel tanks installed.

On June 13, 1979, the first German-produced BK 117 prototype made its first flight, with a Japanese model flying in August. On December 9, 1982, the BK received its German airworthiness certification, and, on December 17, the Japanese Civil Aeronautics Board certified the helicopter. Following these certifications the partnership placed the aircraft into full production, establishing both German and Japanese production lines. Each production facility manufactured individual components, but factories in Donauworth, Germany, and Gifu, Japan, completed final assembly of the helicopter. Deliveries of the BK 117A1 began in 1983, with the improved 117A3 appearing in 1985. In 1987 the 117A4 went into production with a more efficient tailrotor and a transmission that permitted more take-off power. Other improved versions followed, including the German-developed 117C1 with Turboméca Arriel engines and an automatic flight control system. By 2002 combined production in Germany, including Eurocopter, and Japan totaled almost 370 aircraft. The 117 became extremely popular with EMS operators, several police forces, and the German Bundesgrenzschutz (border police).

COMBAT SERVICE SUPPORT HELICOPTERS

In the late 1970s the U.S. Army determined that the Boeing CH-47C no longer met its medium lift requirements. As a result Boeing Vertol modified an existing CH-47A as the prototype for a much-improved CH-47D. In May 1979 the prototype took off on its first flight, and in February 1983 the 101st Airborne Division at Fort Campbell, Kentucky, accepted delivery of the first fully modified CH-47D. In April 1985, impressed with the CH-47D's performance, the Army contracted Boeing Vertol to upgrade all existing "A," "B," and "C" model Chinooks to CH-47D standard.

In phases the Army delivered the Chinooks to Boeing's Ridley, Pennsylvania, facility, where the aircraft were stripped to the airframe and rebuilt. The plant installed two Avco-Lycoming T55-L-712 turbines, more powerful transmissions, and redesigned composite rotor blades. The refit included computerized flight controls, a new hydraulic system, triple cargo hooks, an improved Auxiliary Power Unit (APU), and a single-point refueling. A dual electronic system powered a completely redesigned instrumentation and avionics suite. From the mid-1980s through the mid-1990s, Boeing transformed all CH-47 airframes into either CH-47D or MH-47E aircraft. Boeing rebuilt and upgraded five Chinooks for Argentina, twelve for Australia, ten for Spain, and nine for Canada, which were later sold to The Netherlands. The company also manufactured new aircraft for the U.S. Army, Australia, and Singapore. Spain and Australia initially received CH-47C aircraft, but shortly after delivery of these machines Boeing Vertol introduced the Model 414 International Chinook, which was quickly upgraded to the Model 414-100, basically a CH-47D.

Apart from the United States, the Royal Air Force ordered and used more Chinooks than any other military. In 1978 the RAF ordered thirty-three aircraft to replace its obsolete Bristol Belvederes. Boeing equipped these HC-1s with T55-L-1 IE turboshafts, the latest avionics, rotor brakes, three cargo hooks, provisions for pressurized refueling, and modifications for internal ferry tanks. In the 1982 Falklands War the Chinooks performed so magnificently that the RAF subsequently ordered another eight Chinooks, which included three to replace those lost on the *Atlantic Conveyor* support ship. Despite harsh winds and the most miserable weather conditions, the British crews flew their Chinooks day and night, transporting troops, critical supplies, and the badly wounded to hospital ships. Subsequently the Boeing factory upgraded all the RAF Chinooks to CH-47 specifications, designated the HC-2. The RAF also ordered an additional seventeen new aircraft.

In November 1978, Boeing Vertol announced the Model 234, a civilian variant of the Chinook. In August 1980 company test pilots flew the first prototype of four different models offered. The standard airframe accommodated forty-four passengers, but range varied between the 234LR (Long Range), 234ER (Extended Range), 234UT (Utility), and 234 Combi because of supplemental fuel tanks. Boeing installed airline-type windows, a galley, a small toilet, and enlarged external fairings containing the fuel tanks and baggage compartments along the lower fuselage of the BV-234. British Airways Helicopters bought several aircraft for North Sea oil support and carried more than 80,000 passengers in their first year of operation. Columbia Helicopters, ERA Helicopters, and Helicopter Service A.S. of Norway also bought versions of the Chinook. Unfortunately, passengers generally preferred the smaller Puma and Sikorsky S-61N, which led to a relatively short time of service for the Chinooks in North Sea operations.

On March 1, 1974, Sikorsky test pilots lifted off in the most powerful helicopter produced in the Western world. Based on the CH-53D, three General Electric T64-GE-416A 4,380-horsepower engines powered the CH-53E, which included a new transmission, rotor head, a 79-foot-diameter seven-bladed, titanium-sparred main rotor, and a canted four-bladed tailrotor. The state-of-the-art NVG-compatible cockpit included a Dual Digital Automatic Flight Control System (DDAFCS) coupled to a GPS navigation system. A deicing system gave the aircraft all-weather capability. The Super Sea Stallion could be configured for fifty-five troops or twenty-four litters in its 73-foot, 4-inch fuselage. The helicopter easily handled a 105-mm howitzer, Light Armored Vehicle (LAV), or Hawk SAM system internally, or an external load of 36,000 pounds on its two-point hook system. The helicopter delivered a maximum speed of 170 knots, a service ceiling of 18,500 feet, and a maximum gross weight of 73,500 pounds. A range of 480 nautical miles increased to 990 nautical miles with ferry tanks. With a crew of two pilots, two enlisted flight engineers, and two aerial gunners, the USMC employed the CH-53E aboard all its helicopter assault ships. Sikorsky modified an S-65C as a commercial forty-four-passenger transport, with airstair doors, passenger windows, a lengthened nose, and increased fuel capacity, but it found no buyers.

The MH-53J Pave Low III helicopter is the largest and most powerful helicopter in the USAF inventory. Its technologically advanced systems include a terrain-following, terrain-avoidance radar, FLIR, and a computer-generated cockpit map display, linked to a GPS. The information provided on the pilots' MFD enable the crew to follow terrain contours and avoid obstacles, making low-level, all-weather navigation possible. The USAF utilized the CH-53E for special operations and long-range SAR.

On December 23, 1981, Sikorsky produced the latest in the CH-53 series, the MH-53E Sea Dragon, Airborne Mine Counter Measure (AMCM) helicopter. Sikorsky engineers provided the Sea Dragon with an additional 1,000 gallons of fuel, along with a modified electrical and hydraulic system to accommodate the minesweeping equipment towed behind the aircraft on a floating sled. Additional Sea Dragon duties include shipboard replenishment with both internal and external cargo and SAR missions. All CH-53E variants have in-flight refueling probes and hover in-flight refueling (HIFR) capabilities. Sikorsky installed the most advanced countermeasures to protect the CH-53E in combat, including the APR-39 Radar Warning System, AAR-47 Missile Warning System, ALE-39 chaff and flare dispenser, and the ALQ-157 IR jammer. The U.S. Navy, Marines, and Air Force purchased 175 CH-53Es.

WARSAW PACT

Attack Helicopters

As a result of the success of U.S. helicopter gunships in Vietnam, other nations, especially the USSR, realized the need for armed helicopters. Soviet military doctrine, however, had no place for a helicopter dedicated specifically to the gunship role. In the late 1950s and early 1960s the USSR had armed the Mi-8 Hip and its export version, the Mi-17, but the Red Air Force demanded a fast, heavily armed helicopter to fill the role of the Sturmovik ground support fighter of World War II, or an airborne equivalent of a main battle tank. On September 19, 1969, the Mil Bureau responded with the prototype Mi-24 Hind A.

To produce the first model Hind A (NATO designation), Mil modified the fuselage of the Mi-8 but used the same two TV2-117 1,482-horsepower turboshafts and five-bladed main and three-bladed tailrotor system of the Hip. Mil installed the retractable tricycle landing gear of the Mi-14 and the antidihedral wings of the Mi-6 for weapon installation. The Hind's cockpit that went into flight testing in 1970 resembled a World War II bomber, with a multipaned canopy and a 12.7-mm machine gun in the nose. The pilots sat side by side on a four-place bench seat behind the gunner's position, which resulted in poor visibility. The Hind A carried a crew of three and up to eight combat-loaded troops, who could fire their individual weapons through windows in the cargo compartment. Weapons on the wing stores included four to eight AT-2 Swatter AT-GMs and two to four 57-mm rocket pods. If not transporting troops the Hind held four litters and a medic, or carried a second basic load of rockets and missiles internally. The heavily armored Hind A, according to the Mil Bureau, posted a speed record of 198.72 knots during testing. The West first saw the Hind A in Eastern Europe in 1972, with "V" and "C" models appearing in succeeding years.

In 1975, Western intelligence services discovered the radically redesigned Mi-24D. A new stepped tandem cockpit with bulletproof bubble canopies provided greater visibility for the pilot and copilot/gunner, who sat in the forward cockpit, just behind and above a YaKB-12 four-barreled heavy machine gun mounted in a chin turret capable of a 120-degree traverse. Two Isotov TV-3-117 2,200-horsepower turbines, installed in the upper section of the 57-foot, 8-inch fuselage, powered the all-metal 56-foot, 9-inch main and 12-foot, 9.5-inch tailrotors. With a wingspan of 21 feet, 4 inches, the new Hind exhibited a range of 245 nautical miles with a normal load, and a maximum ceiling of 14,700 feet at a maximum gross weight of 26,455 pounds. Without weapons the aircraft could haul a 5,500-pound slingload. In addition to armored seats, applique armor surrounded the cockpit as well as critical oil and fuel supplies. Wingstores included ATGMs, 57- or 80-mm rocket pods, or free-fall bombs.

The Mi24D began to appear in significant numbers in Soviet units in 1976, and in Warsaw Pact countries shortly afterward. Production records indicated that about fifteen Hinds a month rolled off the Mil assembly lines. At the time the Red Army invaded Afghanistan in 1980, more than 1,000 Mi-24s were in service, and the Hind became a symbol of that war, much like the Huey in Vietnam. Although Mil upgraded the Hind with lighter, more efficient composite rotor blades, yokes, and hubs, aircraft limitations affected the successful employment of the Hind in the rarified air of the Afghan mountains. The wings provided 22 to 28 percent of the helicopter's lift, requiring pilots to maintain minimum forward airspeeds or the helicopter would experience unmanageable roll rates in tight turns; nor could the heavily loaded machines hover at the high altitudes encountered, sometimes 18,000 feet. Although ruggedly designed, the Hind's transmission, and especially the tailrotor gearbox, rapidly overheated at a hover without the cooling effect of airflow through cooling vents. As a result the Hind pilots mimicked U.S. Army tactics from Vietnam and flew in pairs, or multiples of pairs, making running fire attacks on their mujahideen adversaries. The Hind pilots relied on speed and armor to survive. They attacked at 140 knots, blasted the target area, and pulled away in tight turns.

Soviet tactics overall replicated U.S. tactics in Vietnam. Mi-8 and Mi-17s, escorted by Mi-24s, lifted large numbers of troops to air assault into remote areas to attack mujahideen soldiers in their sanctuaries. Hind pilots also frequently flew "roadrunner missions," escorting vulnerable convoys moving along winding mountain roads. Afghani rebels called the Mi-24 the "Devil's Chariot" because of the heavy firepower the Hind brought to the battlefield. The Hind pilots called themselves "Grey Wolves."

Ground fire downed several other types of helicopters, but the heavily armored Mi-24s remained almost impervious to most weapons, except Rocket Propelled Grenades (RPGs). To escape the volleys of RPGs most Soviet helicopter crews flew at higher altitude until 1985, when the CIA introduced the U.S.-manufactured Stinger missiles through Pakistan. The highly effective Stinger, with a maximum range of 15,000 feet, forced the helicopters back down where small arms again began to take a toll of Soviet aircraft. The rebels claimed that all they needed to defeat the invaders was the Koran and more Stingers. The USSR lost hundreds of aircraft and at least 15,000 aircrewmen in the Afghanistan War. The mujahideen claimed to have downed more than 200 Mi-24s alone. Several captured Hind crews were skinned alive because of the death and destruction they wrought on rebel villages. In 1987, Soviet engineers equipped their helicopters with flare dispensers, but the Stingers continued to bring down helicopters until the last Soviets departed Afghanistan in February 1989.

On November 10, 1982, the Mil OKB began testing the Mi-28 Havoc, intended to replace the Hind. With its stepped, two-place tandem cockpit, two Klimov TV-3-117VM, 2,200-horsepower turboshafts installed externally on either side of a long, slim fuselage, and a large tailfin mounting an asymmetrical X-shaped tailrotor, the Mi-28 bore a great resemblance to the AH-64 Apache. A nose radome housed a laser rangefinder and radar. A 56-foot, 5-inch five-bladed composite main rotor provided lift for Mil's new attack helicopter. Smaller than the Mi24 at a maximum gross weight of 24,500 pounds, the Havoc, nonetheless, packed a significant wallop. Typical armament included a chin turret mounting an A42 30-mm cannon, sixteen AT-6 or AT-9 ATGMs, and forty S-8 rockets or two GSh-23 23-mm cannons on the stub wings. The wings also held ECM pods at their tips. For some time the Havoc created quite a stir among Western intelligence operatives, as well as helicopter pilots, but the Mi-28 failed to live up to its hype of a fully aerobatic attack helicopter. Never placed in full production, the Mi-28's maximum speed appeared to be around 160 knots and its range 250 nautical miles. In the mid-1990s Mil introduced the Mi-28N with a mast-mounted FLIR for enhanced night operations, but the Russian military seemed inclined toward the Kamov Ka-50 Hokum as its primary attack helicopter. To date the Mil bureau has not been able to find foreign customers for the Havoc.

In 1982 a prototype of a revolutionary Soviet attack helicopter appeared that sent chills through most NATO helicopter pilots. The Kamov Ka-50A Blackshark, designated Hokum by NATO, looked as much like a single-seat jet fighter as it did a helicopter. Although U.S. Army generals denied any necessity for air-to-air capabilities in Army helicopters, and USAF generals promised protection from all low-flying aircraft, the Ka-50A negated both assumptions. Two Klimov TV3-117VMA 2,200-horsepower turboshafts, installed on either side of the slim 44-foot, 3-inch fuselage just above the wing-roots, powered the 45-foot, 7-inch three-bladed, swept-tipped polymeric coaxial rotors, which also incorporated an electric deicing system. The fuselage, constructed of more than one-third composites, including a kevlar/nomex armored keel, ended in a fixed-wing type empennage and held the retractable tricycle landing gear. IR suppressers covered the engine exhausts, and OKB equipped the aircraft with IR jammers, radar warning receivers, and chaff/flare dispensers. A fully armored seat protected the pilot from 23-mm rounds, and the flat-plate canopy deflected anything up to 12.7-mm fire. A Zvezda K-37-800 pilot ejection system allowed the pilot to eject from the Ka-50 at low airspeeds and altitudes. Explosive bolts separated the rotor blades from the bearingless hub at the initiation of the ejection sequence.

Designed as an antitank/antihelicopter aircraft, the fully aerobatic Hokum carried a variety of weapon systems. Acquisition and targeting systems included low-light television and laser rangefinders/designators linked to a satellite navigation system and automatic pilot that allowed the Hokum pilot to engage targets at ranges over 10 kilometers. A helmet sighting system and heads-up display (HUD) allowed the pilot to focus his attention outside the cockpit while flying in adverse conditions or operating the weapons systems. The fire control computers allowed the pilot to engage targets outside his visual range, and a digital downlink provided the target data to a ground control center. For day/night, all-weather operations the Ka-50N, sometimes called the Nightshark, or more popularly Werewolf, carried a nose-mounted FLIR and millimeter-wave radar in an EO (electro-optic) underwing pod, and the cockpit had an additional MFD. Capable of carrying more than 5,000 pounds of ordnance on the wingstores, the Ka-50 could be armed with up to sixteen AT-9

Vikhr antitank missiles, with two 20-round S-8 80-mm FFAR pods, and 500 rounds, mixed HE and AP, for the 2A42 30-mm cannon, the same gun mounted on the BMP-2. The enhanced version of the supersonic 125-mm Vikhr missile depended on radar guidance during launch and laser guidance for target designation. The two-stage shaped-charge warhead penetrated armor up to 900 mm.

Making the AT-9 even more deadly, the Ka-50 pilot, by a flick of a switch, could engage aircraft flying at up to 450 knots with the AT-9. Twin 23-mm cannon pods, AS-12 Kegler guided missiles, AA-11 Archer and IGLA-V, Needle C, AAMs, and 1,000-pound bombs also appeared on test aircraft. The Ka-50A attained a known speed of 188 knots and reportedly reached a maximum range of 650 nautical miles with auxiliary fuel tanks, and 240 nautical miles with maximum ordnance load. Reported service ceiling was just over 18,000 feet. With the demise of the USSR, the Hokum failed to reach full production by 2000, but the Russian Air Force intended to acquire two aircraft per year for fourteen years, depending on available funding.

Naval Helicopters

In 1973 the Soviet Aviation Ministry issued directives to develop an attack/assault transport helicopter for support of naval infantry and amphibious operations. OKB Kamov's Deputy Chief Designer S. N. Fomin led the program with leading designer G. M. Danilochkin and leading engineer B. V. Barshevsky as his chief assistants. On July 28, 1976, test pilot Y. I. Laryushin lifted the Ka-29 prototype off on its first flight. The design bureau completed all acceptance trials by May 1979 and placed the Ka-29 Helix B in full production in 1984.

Based on the Ka-27 Helix, OKB widened the fuselage and revamped the forward section with a five-piece flat windscreen and blunt nose, which housed a FLIR/TV sighting system and a new search/targeting radar. Armament stations included a fixed multiple-barreled 7.62-mm machine gun under the right side fuselage, and winglets on which to mount a variety of weapons. Two Klimov (Isotov) TV3-117V 2,190-horsepower turboshafts turned two typical Kamov three-bladed 52-foot, 2-inch coaxial rotors, which allowed the Ka-29 to take off at a maximum gross weight of 27,775 pounds. This translated into two pilots and up to sixteen combat-loaded troops, or four litters and six seated patients with two attendants in the air ambulance modification, or an 8,800-pound slingload. Typical weapons loaded on the Ka-29TB attack version included four 57- or 80-mm rocket pods, or two rocket pods and two four-round clusters of AT-6 Spiral ASMs. In addition to a 30-mm cannon mounted above the left wing, the helicopter could also be armed with submunition dispensers (CBUs) or conventional free-fall bombs. In several comparison tests with the Mi-24D Hind, the Ka-29TB, because of the almost vibrationless rotor system, proved almost twice as effective at placing its ordnance on target as the Hinds.

The pilots enjoyed the communications and electronics suite provided in the new Helix B. These systems included a Doppler radar, and later GPS, navigational system, integrated with computerized displays of flight and targeting information incorporated into a modern cockpit layout. All versions cruised at 125 knots with a maximum airspeed of 151 knots, and a maximum range of 400 nautical miles. The Soviet Navy planned on a combat radius of 54 nautical miles, including six to eight attack passes for the Ka-29TB.

In the early 1980s the Soviet Union provided ASW helicopters to other countries. With the advent of the Ka-29 the USSR sold the Ka-28, a downgraded export version of the Ka-27, to India, Ukraine, and the Socialist Republic of Vietnam. The Ka-28 carried a dipping sonar, disposable sonobuoys, and wire-guided torpedoes, or depth charges, but not the latest in electronic submarine detection gear. The Soviets sold their allies an upgraded version of the equipment carried by the Ka-27, but not the advanced electronics installed on the Ka-29 Helix.

On October 8, 1980, a prototype medium lift multipurpose version of the Ka-27 also appeared. Intended as a commercial helicopter and known as the Ka-32 Helix C, it had two Klimov TV3117V 2,190-horsepower turboshafts that turned the same counter-rotating three-blade main rotors installed on the Ka-29. The several versions of the Ka-32 also had the wider fuselage of the Ka-29, indicating a probable developmental link between the two machines.

A pilot and navigator crewed the Ka-32T transport version, which accommodated sixteen passengers, or an internal load of 8,820 pounds, or an 11,000-pound slingload. The Helix C appeared in passenger/cargo transport, air ambulance, fire-fighting, police, flying crane, and SAR versions. The Ka32K featured a retractable underfuselage gondola for a second pilot to fly the aircraft while picking up or delivering bulky slingloads. The Ka-32S SAR helicopter included a search radar, as well as advanced flight and navigation instrumentation for IFR and maritime operations. The Russian government and commercial operators also made use of the Ka-32S in offshore oil explorations. Without a slingload the Ka-132 attained a maximum airspeed of 135 knots and a range of 430 nautical miles without auxiliary fuel. Although described as a commercial helicopter, and sold or leased to several foreign countries, Ka-32s in Aeroflot colors were photographed operating from the decks of vessels belonging to the Russian Navy.

As the economy of the USSR decayed, Soviet, then Russian, industries began to seek civil and foreign markets for their products. In October 1988, Kamov introduced the first of fifteen civilian variants of the Ka-126 derived from the naval Ka-26 Hoodlum. The Ka-126 featured a modular concept to rapidly convert the light, multipurpose helicopter to accomplish several diversified missions. Wide use of composites in both the traditional three-bladed coaxial rotors and fuselage lightened the aircraft, which resulted in increased load capacity and range. A single TVO-100 720-horsepower turboshaft, mounted above the cabin, provided power to lift a pilot and six passengers, or an internal cargo load of 2,200 pounds. The 126 cruised at 90 knots and attained a service ceiling of 15,250 feet. Kamov intended the Ka-126 to fulfill EMS, police, passenger/cargo transport, and geological/oil survey roles. The agricultural version, designed especially for crop spraying, was equipped with a cockpit air filtration system to prevent toxic chemicals from entering the flight deck. Kamov installed a 722-horsepower Turbomeca Arriel 1D1 turboshaft in one export version of the helicopter.

Combat Service Support

The Mil design bureau holds the distinction of designing and building the largest helicopter placed into full production. The Mi-26 Halo, with a maximum gross weight of 123,650 pounds, corresponds in size to a Boeing 737. Designed as a heavy lift military transport to replace the Mi-12, the huge machine was capable of hauling a 45,000-pound payload or seventy fully equipped troops. The Halo, however, became most successful as a civilian helicopter, earning fame by resupplying remote Siberian villages and oil camps; fighting forest fires throughout the world; and providing a mobile crane for construction of high-rise buildings, bridges, or pipelines in remote areas.

First flown on December 14, 1977, the Halo had an aerodynamic pod and boom fuselage that measured 131 feet, 4 inches in length, with a spacious cockpit for the crew of four forward, and large clamshell doors aft. The Halo lacked the wings of the Mi-12, depending on an advanced rotor design for all its lift. Powered by two Lotarev D-136 5,620-horsepower turboshafts mounted atop the fuselage, driving a 104-foot, 11.5-inch eight-bladed main and five-bladed composite tailrotors, the Halo reached a service ceiling of 15,000 feet. In a clean configuration, without an external load, the Mi-26 was capable of a maximum speed of 160 knots, usually cruising at 135 knots, with a normal range of 360 nautical miles. With internal ferry tanks the range increased to 1,100 miles, permitting the big machine to self-deploy over long distances. The helicopter rested on a very robust fixed tricycle landing gear. Mil produced at least 550 Mi-26s, improving the machine's performance and versatility, with the most current variants incorporating engines up to 8,500 horsepower each, more efficient rotor systems, and digitized glass cockpits. The MJ-26 boasted a 100-troop capacity, improved rotor blades, and a flight director with an autohover mode. The Mi-26T "flying crane" included a modified flight deck with a second pilot position and stabilization system for lifting and depositing cumbersome external loads. In a firefighting role the Mi-26 carried up to 4,400 gallons of water in two large buckets. India bought twenty Mi-26 export versions, and Ukrainian Haloes, under UN colors, served in Bosnia. Several countries used the Halo on a contract basis for construction projects and for fighting large fires.

Compound Aircraft, Convertaplanes, and Tilt Rotors

In the late 1940s and early 1950s, just as the helicopter was proving its practicality, aviation engineers initiated design programs to produce a machine that incorporated the best characteristics of both fixed- and rotary-wing aircraft. One of the first began just at the close of World War II. In 1945 several employees left Piasecki Aircraft and formed the Transcendental Aircraft Company in order to investigate tilting rotor concepts. In 1951 the company introduced its first rotorcraft, the single-seat, open-cockpit Model 1G. Like all pioneering tilt rotor machines, the 1G suffered from dynamic instability linked to the rotor systems. The Transcendental Aircraft Company requested, and received, funding from the USAF to research gyroscopic effects and airflow unique to rotorcraft. After several modifications the 1G lifted off vertically on July 6, 1954, and made its first conversion to horizontal flight the following December. The

26-foot fuselage housed a single Lycoming O-290-A 160-horse-power piston engine, which powered the three-bladed wingtip rotors through a series of gearboxes and driveshafts. Three concentric shafts converted the pilot's control inputs to tilt the rotors and change cyclic and collective pitch. A complete transition of the rotors required a full three minutes. The 1,750-pound 1G measured 21 feet from wing tip to wing tip and posted a top speed of about 160 mph. Before a control malfunction caused a crash on July 20, 1955, the 1G flew more than twenty hours in over 100 flights. In 1956 and 1957 the company tested a heavier, 4,000-pound model, powered by a 250-horsepower engine, but the USAF dropped the program in favor of the Bell XV-3.

The next U.S. tilt rotor project took shape in 1949, when the McDonnell Aircraft Corporation expanded a design concept from Austrian aviation pioneer Friedrich von Doblhoff and modified the company Model 82 into what became the XV-1. On June 20, 1951, McDonnell, the USAF Air Research and Development Command, and the U.S. Army Transportation Command signed a letter of intent to fund and participate jointly in developing an experimental vertical lift aircraft based on the rotor principle. In this principle, when the aircraft accelerates past stall speed, the rotor goes to flat pitch and "windmills" like an autogyro. Actually, one of three concepts including the Bell Model 200 XV-3 Tilt-Rotor and the Sikorsky S-57/XV-2 Retractable Rotor vehicle, the XV-1, flew first, and the Sikorsky design never left the drawing board.

Based on a twin-boomed, twin-tailed fuselage, much like the Lockheed P-38 Lightning, the XV-1 incorporated a 30-foot, 6-inch three-bladed main rotor for vertical ascent/descent and a pusher propeller for forward thrust. A 29-foot pod centered on the 25.5-foot wing housed both a four-place glass-enclosed cockpit in the forward section and the engine and transmission aft. A fixed skid landing gear beneath the pod completed the resemblance to a helicopter. A Continental R-975-42 525-horsepower reciprocating engine powered both propeller systems through a complicated drive system. The radial engine powered two compressors that sent ducted air from the rotor hub to pressure jets at the blade tips, providing vertical lift. When the aircraft reached flying speed, the pilot transferred engine power to the two-bladed pusher prop. On February 11, 1954, McDonnell test pilot John R. Noll began a series of tethered hovering flights, but pressure problems in the main rotor precluded free flights until July 14. On April 29, 1955, the XV-1 made its first successful transition from helicopter to fixed wing flight.

A second, more streamlined prototype featured improvements, including additional directional rotors at the aft ends of the tailbooms and more robust skids. Designed to carry a pilot and copilot in side-by-side configuration, or a pilot and three passengers, the XV-1 could also be configured with two litters and a pilot. Flight records indicated cruise airspeed of about 100 knots, although, on October 10, 1956, the second prototype became the first rotorcraft to reach an airspeed of 173.88 knots The craft weighed 4,750 pounds at maximum gross weight and reached an altitude of 11,500 feet. The XV-1 did not achieve great success compared with conventional helicopters, and the piston engine provided insufficient power. Engineers determined that a turbine engine would improve the aircraft's performance, but the program was terminated in 1957, after the two prototypes had accumulated more than 600 hours of flight time.

Even after the success of the Model 47, Larry Bell and his engineers realized that a helicopter could never compete with certain capabilities of a conventional airplane. As a result Bell designers conceptualized a machine that would take off and land like a helicopter but could tilt its wings and rotors and fly like an airplane. In 1950, following a design by Rol Lichten, Bell Helicopter initiated the Model 200 project. The previously mentioned June 1951 letter of intent included two prototypes of each convertaplane. Bell began constructing the first XV-3, which the company unveiled on February 10, 1955.

On August 11, 1955, Chief Pilot Floyd Carlson lifted off in the all-metal, four-place XV-3 on its first vertical flight. A single Pratt & Whitney R-985 450-horsepower radial engine, mounted behind the Plexiglas cockpit, turned the two 33-foot, 10-inch fully articulated, three-bladed rotors mounted at the tips of the 31-foot, 3.5-inch cantilevered wings. A series of transmissions, gearboxes, and driveshafts transferred power from the engine to the prop/rotors, which were rotated by electric motors. The 30-foot, 3.5-inch fuselage included conventional tail surfaces and a twin-skid landing gear. With the prop/rotors tilted up, the XV-3 lifted off like a helicopter. Then, when they were tilted forward, the prop/rotors powered the aircraft in forward flight, with the fixed wing providing the necessary lift. Despite exhibiting some instability, the first prototype achieved several partial transitions before an August 23, 1956, accident seriously injured the pilot, Richard "Dick" Stansbury, grounding both Stansbury and the aircraft.

In late 1957, Bell engineers installed two 24-foot, two-bladed semirigid rotors on the second prototype. On December 18, 1958, the second XV-3 made its first full transition from vertical to horizontal flight. The XV-3 demonstrated a range of airspeeds from 15 mph backward to over 180 mph in forward flight, at altitudes up to 12,000 feet, with a maximum takeoff weight of 4,800 pounds. In 1962 a prop/rotor pylon instability problem grounded the second prototype after 110 successful full-transition flights. In 1965, Bell Helicopter transferred the XV-3 to NASA, which utilized the aircraft in wind tunnel tests until it was irreparably damaged. NASA then discarded the machine.

On April 15, 1956, the Vertol Aircraft Company entered into a joint U.S. Army and Navy program to pursue the company Model 76 design as the VZ-2A. Viewed from the front the aircraft resembled a helicopter, with its 26-foot, 6-inch tubular frame fuselage and large bubble canopy surrounding the two-place cockpit, but the short wings and tall "T-tail" belied that conception. A 660-horsepower Lycoming YT53-L-1 turboshaft, mounted above the fuselage, through a complex series of shafts and gearboxes, powered a set of ducted fans on the tail and two counter-rotating 9-foot, 6-inch three-bladed propellers mounted near the center of each wing. The variable-pitch ducted fans installed horizontally and vertically on the tail structure provided pitch and yaw control at a hover, while the large propellers allowed the aircraft to lift off as a helicopter and then transition into forward flight. The VZ-2A's wings rotated from the vertical to the horizontal, allowing the machine to accelerate forward, where aerodynamic forces on the traditional wing and tail surfaces provided lift to sustain flight.

In April 1957 the VZ-2A made its first vertical takeoff, with several more hovering flights occurring during the next year. On July 15, 1958, the aircraft made its first complete transitional flight, lifting off to a hover, transitioning to cruise flight, and returning to land vertically. Although highly maneuverable, the 2,700-pound rotorcraft reached an airspeed of only 116 knots, even after the fuselage was covered with a thin aluminum skin to improve its aerodynamics. NASA acquired the aircraft and continued to fly it until 1965. In all, the VZ-2 made 450 flights, including 34 full conversions, proving the tilt-wing a feasible concept.

In 1973, Bell Helicopter concluded negotiations with both the U.S. Army Air Mobility Research and Development Laboratory and NASA to become the prime contractor for a project researching the viability of a tilt rotor aircraft. Using the Model 200 as a data base, Bell designed the Model 301. With additional funding Bell produced the first of two prototype tilt rotor aircraft, designated the XV-

15 by NASA, the first of which flew on May 3, 1977. After its initial flights NASA relegated the first prototype to wind tunnel testing. In April 1979 the second prototype made its first flight and successfully transitioned from vertical to horizontal flight on July 24.

The XV-15's streamlined fuselage, upswept tail with twin vertical fins, and tricycle landing gear resembled those of a conventional airplane, but the shoulder-mounted cantilever wings, with the large engine nacelles and 25-foot, stainless-steel prop-rotors at the tips, bespoke a revolutionary aircraft design. Each nacelle housed an Avco Lycoming LTC1K-4K 1,550-horsepower turboshaft, with transverse cross-shafting, permitting either engine to drive both proprotors in the event of a single engine failure. The cockpit accommodated a pilot and passenger, or two pilots, in side-by-side seating. By manipulating conventional controls the pilot could lift off vertically and hover the XV-15 like a helicopter, then tilt the engine nacelles forward to translate into horizontal flight. The XV-15 reached a maximum airspeed of 332 knots, a service ceiling of 29,000 feet, and achieved a maximum takeoff weight of 13,000 pounds. The aircraft recorded a maximum range of 445 nautical miles with a standard fuel load, surpassing all helicopters with a combination of speed, range, and cargo potential. The success of the XV-15 led to a joint Bell-Boeing venture to compete for the Joint Services Advanced Vertical Lift Aircraft (JVX) program. The military designated the Bell-Boeing Model 901 the V-22 Osprey.

The Curtiss-Wright Company produced two aircraft making use of tilting propellers instead of rotors. The tilting propeller concept incorporated short, rigid propellers with collective pitch control, but no cyclic, relying on traditional aircraft controls for directional control. The X-100 employed engine exhaust, ducted to the rear of the 24-foot fuselage, to provide pitch and yaw control in hover flight; differential propeller thrust accounted for roll control. A Lycoming YT53-L-1 drove the 10-foot-diameter tilting fiberglass propellers mounted at the tips of the 16-foot wings. The 3,500-pound X-100 completed its first untethered hover in September 1959 and made a short flight the following March. In April 1960 the aircraft completed its only fully transitional flight. Control at a hover was marginal because of low exhaust gas velocities, but testing continued until October 1961, encouraging Curtiss-Wright executives to proceed with the company's next project.

Using the X-100 as a basis, Curtiss-Wright engineers designed the X-200, a six-passenger executive transport. The USAF funded the conversion of two prototypes for an assault transport, designated

X-19. The USAF demanded numerous modifications to meet military requirements, including ejection seats, a rescue hoist, a refueling probe, and a lengthened fuselage to improve passenger access. Two Lycoming T55-L-7 2,650-horsepower turboshafts drove 13-foot, three-bladed wide chord wingtip propellers. Differential propeller thrust controlled roll, pitch, and yaw. On November 20, 1963, the first prototype completed its initial hover flight but suffered a hard landing. Although the X-19 was repaired, a series of control problems and mechanical malfunctions plagued the craft. On August 25, 1965, a transmission failure caused the crew to eject from the stricken aircraft. Four months later the USAF terminated the X-19 program. Although the first prototype had made fifty flights, its time in the air totaled only four hours. The company never flew the second prototype.

Throughout the 1950s, 1960s, and 1970s, many other U.S. aircraft manufacturers experimented with V-STOL (Vertical-Short Takeoff and Landing) and other unique aircraft designs, many dismissed as mere contraptions. DoD funded many of these radical projects in an attempt to find a machine designed to move one, two, or several individuals around the modern battlefield. Hiller Aircraft produced the VZ-1, "Flying Platform," powered by a small, air-cooled piston engine. About 6 feet in diameter, the platform was designed to carry a single soldier across nuclear contaminated areas, or over battlefield obstacles.

In February 1957 the Hiller Helicopter company cobbled together components from several aircraft to construct the X-18. Two Allison T40-A-14 7,100-horsepower turboshafts, mounted on the wings, drove the 16-foot, counter-rotating, three-bladed propellers, and a Westinghouse J34 turbojet, which produced 3,400 pounds of thrust, provided forward momentum. Beginning in December 1958 the 33,000-pound X-18 underwent a series of rigorous ground tests. On November 24, 1959, the aircraft completed its first conventional flight. The electric pitch controls on the turbine engines proved too unresponsive, however, to provide adequate control in hovering flight. During the X-18's twentieth flight a propeller pitch failure at 10,000 feet induced a spin, but the pilot recovered control before impact. The incident, however, grounded the machine before it achieved a successful hover.

When Fairchild Aircraft acquired the Hiller Aircraft Company in the mid-1960s, the company introduced the M-224-1/VZ-5 Fledgling to meet a U.S. Army requirement for an experimental tilt-wing aircraft. A single General Electric YT58-GE-2 1,024-horsepower turboshaft turned four three-bladed propellers, and two small horizontal rotors at the top of the T-tail, which controlled pitch at a hover. In 1959 the VZ-5, a two-place 4,000-pound rotorcraft, lifted off in tethered flights, but it never met its designed speed of 160 knots in the airplane configuration. Although the Fledgling made several successful flights, the program did not progress beyond the test stage.

In June 1956 the Ryan Aircraft Company introduced the Model 92, designated VZ-3 by the Army, intended to be a reconnaissance and liaison aircraft. The 28-foot metal fuselage housed a 1,000-horsepower Lycoming T53-L-1 turboshaft, powering two three-blade propellers, mounted well forward and below the wings. The majority of the propeller wash was directed into extended flaps, turned downward to provide vertical lift. Engine exhaust, ducted to the rear, furnished pitch and yaw control at a hover. On February 7, 1958, Ryan began taxiing trials, and, after several modifications, the VZ-3 completed its maiden flight on January 21, 1959. Unfortunately the engine proved underpowered, and the aircraft could not hover except in a headwind. The VZ-3 suffered several accidents, resulting in further modifications and relegation to low-speed testing by NASA in 1961.

Charles Kaman produced a prototype tilt-wing aircraft based on a Grumman Goose amphibian's fuselage and a tilting wing. Two General Electric T58-GE-2A turboprop engines powered the unusual machine. In 1978, Kaman introduced an experimental compound helicopter, with short-span wings and a conventional tail. The aircraft carried several modifications for high-speed flight, including a GE YJ85 turbojet engine installed on the starboard fuselage. The U.S. Army evaluated the aircraft but expressed no interest in pursuing the project past the prototype machine. On December 29, 1971, the Kaman "Saver" made its first flight. The diminutive autogyro, a foldable, single-seat, emergency aircrew escape aircraft with telescoping rotor blades, generated some interest in the U.S. military but produced no orders. A single Williams Research WR-19 turbofan engine provided power for the little rotorcraft, intended for downed pilots to fly themselves out of enemy territory.

As a result of a 1959 recommendation from a government advisory panel, the Vought-Hiller-Ryan Group, later known as Ling-Temco-Vought (LTV), produced five prototypes of the XH-142 combat assault aircraft. Designed to carry thirty-two fully equipped troops, the tilt-wing aircraft made its first flight on September 29, 1964, and on January 11 of the following year it completed its first fully transitional flight. Four General Electric T64-GE-1 2,850-horsepower turboshafts each powered a 15.5-foot four-bladed propeller. A series of shafts and gears also linked the engines in the event of an engine failure, and to drive the 8-foot three-bladed tailrotor. The wing tilted to 100 degrees vertically, allowing the machine to hover in a tailwind. The XH-142 measured 58 feet in length and had a wingspan of 67 feet. A rear cargo ramp simplified the loading of 8,000 pounds of cargo. With a maximum gross weight of 41,000 pounds for a vertical takeoff, the machine reached a top speed of 431 mph, a range of 820 miles, and a service ceiling of 25,000 feet. Extraordinarily loud and susceptible to numerous vibrations, the XH-142 exhausted its pilots and suffered a number of equipment failures, especially in the complicated arrangement of driveshafts. A catastrophic tailrotor driveshaft failure resulted in a crash killing all three on board. By 1967 only one of the prototypes remained flyable. Although the XC-142 accrued 420 hours by thirty-nine different pilots and completed a range of testing from troop/cargo transport to carrier operations, its propeller thrust, and as a result, its performance proved unsatisfactory; the U.S. military canceled the project. A proposed civilian version of the XH-142, the "Downtowner," intended to carry forty to fifty passengers at a cruise speed of 290 mph, never got off the ground.

The Canadian CL-84 program, begun in November 1963, resulted in an aircraft flown in May 1965. Much smaller than the XC-142, the CL-84 could lift off vertically at 12,200 pounds. Two wing-mounted Lycoming T53-LTC1K-4A 1,450-horsepower turboprops turned the 14-foot, four-bladed propellers. Drive shafts linked the two engines and the two counter-rotating two-bladed horizontal rotors, which provided pitch control during hover flight. To minimize drag the small rotors stopped and aligned with the fuselage during forward flight. Differential propeller pitch compensated for roll, and yaw was controlled with ailerons. In December the rotorcraft completed its first conventional flight. Of the four prototypes constructed, two were destroyed in nonfatal accidents. Although U.S. pilots evaluated the CL-84, including landings on carriers and the Pentagon helipad, neither government ordered production aircraft.

In February 1972, Sikorsky Aircraft announced the development of the first of several experimental aircraft that diverged dramatically from the company's traditional line of helicopters. Designed with two counter-rotating three-bladed rigid rotors, the S-69, designated XH-59A by the U.S. Army, Advanced Blade Concept (ABC) was to determine if the concept led to higher airspeeds and greater agility. Two externally mounted Pratt & Whitney PW J-60 turbojets drove the coaxial rotors, while the rudders on the twin tail provided directional control in flight. With the aerodynamic load placed on the two advancing blades of the coaxial rotors, the streamlined aircraft required no supplemental wings to provide additional lift to attain high speeds. On July 26, 1973, the ABC made its first flight, but it crashed a month later. After several modifications, test flights resumed with a second prototype in July 1975. In 1977, after the conclusion of flights as a pure helicopter, Sikorsky converted the ABC into a compound rotorcraft by installing two J60-P-3A turbojets, which allowed the aircraft to attain level airspeeds in excess of 278 knots. The Army, Navy, and NASA all evaluated the aircraft, and both the second prototype and the rebuilt first model were transferred to NASA in 1981.

The second experimental machine, the S-72 Rotor Systems Research Aircraft (RSRA), received joint funding from NASA, the U.S. Army, and Sikorsky. Designed as a flying testbed for experimental rotor systems, the RSRA flew either as a helicopter or a compound aircraft. The RSRA became unique as the only helicopter ever to have a certificated crew ejection system. The system included an explosive charge to shear the blades from the hub, and then the crew ejected as from a fixed-wing aircraft. The RSRA was later used as a flight test vehicle for the X-wing program.

In 1986, Sikorsky began testing the X-wing aircraft, a Modified S-72 RSRA helicopter. The X-wing employed an unusually shaped composite rotor system to lift off and transition into forward flight. When the aircraft reached a suitable forward speed, the pilot stopped the rotor, which locked in place and became a fixed wing for the aircraft. The X-wing made several successful flights, providing a wealth of aerodynamic data, but it did not result in any production aircraft making use of the X-wing concept. Across the Atlantic Ocean aviation engineers and manufacturers in Europe succeeded in producing compound aircraft, tilting wing, or tilt rotor aircraft that met with varying degrees of success. On May 8, 1953, the French Sud-Ouest SO-1310 Farfadet lifted off on its maiden flight. An all-metal three-seat compound helicopter with small wings and fixed tricycle landing gear, the SO-1310 relied on two engines for propulsion. A Turbomeca Arrius II 360-horsepower turboshaft drove the four-bladed main rotor, and a Turbomeca Artouste II powered the propeller that provided momentum for forward flight. The Farfadet made several successful flights, but Sud-Ouest produced only the single prototype.

In 1966 the Nord Company, later acquired by Aerospatiale, introduced the NORD 500 Cadet. NORD fabricated two prototypes, each constructed around two Allison T63-A-5A 317-horsepower turboshafts. The aircraft measured 22 feet in length, 20 feet wide, and weighed only 2,760 pounds. Two large five-bladed ducted propellers provided both lift and thrust for the Cadet. NORD utilized the first airframe for static tests, while, on July 23, 1968, the second machine made a single tethered, hovering flight before the project was canceled.

On November 6, 1957, the Fairey Rotodyne, the largest compound aircraft of its era, completed its maiden flight. In August 1953 the British Ministry of Aviation, based on the success of the Fairey Gyrodyne prototypes, which had established a number of British helicopter records, ordered the forty-passenger rotorcraft. Two wingtip Napier Eland 2,800-horsepower turboshafts provided pressurized air to turn the large 90-foot rotor system, which in turn lifted the aircraft vertically for takeoffs, landings, and hovering. Conventional tractor turboprops powered the aircraft in forward flight. On April 10, 1958, a Fairey crew completed the first flight using the tractor propellers and, on January 5, 1959, established a helicopter speed record of 165.8 knots. When Westland Aircraft acquired the Fairey company, it proposed to enlarge the Rotodyne to carry fifty-five to seventy passengers, or 14,800 pounds of cargo, including standard British Army vehicles. Westland also installed two Rolls-Royce Tyne 5,250-horsepower turboprops to raise cruising speed to 200 knots. Unfortunately, the RAF canceled an order for twelve machines, and the interest of British European Airways waned, forcing the cancellation of the Rotodyne project in February 1962.

In 1959, attempting to compete with Mil's heavy lift helicopters, a team of Soviet aviation engineers headed by Nikolai I. Kamov designed and built a prototype compound helicopter even larger than the Rotodyne: the Ka-22 Vintokryl, designated Hoop by NATO. Y. S. Braginsky, leading design engineer on the project, forswore Kamov's typical coaxial rotor system for two large four-bladed main rotors. Two Solovyov D-25VK 5,500-horsepower turboshafts, installed in wingtip nacelles, each drove a rotor and a tractor propeller. The high-mounted wings and rotors spanned 152 feet, and the Ka-22 measured 80.5 feet in length. After a vertical takeoff, the pilot transitioned into forward flight, where the wings produced the majority of the aircraft's lift. At this point the pilot transferred engine power to the propellers, and the rotors freewheeled as on a gyroplane. On August 15, 1959, D. K. Yefremov flew the Hoop on its first flight.

Two years later, on October 14, Yefremov and his crew set the first of eight world records established by the Ka-22 by reaching an airspeed of 188 knots. The Hoop also set the payload to height record by lifting 36,350 pounds to 6,500 feet. Flight records listed the Ka-22's maximum gross weight at 93,700 pounds, and a useful load of 36,750 pounds, but with a range of less than 250 nautical miles. After two unexplained crashes in the early 1960s and the success of the more conventional Mil Mi-6, the Soviet Air Ministry dropped the Ka-22.

In 1968, Fuji Industries of Japan ventured into experimentation with compound aircraft. Using a Bell 204B-1, the company fabricated a high-speed compound version of the Huey, called the Fuji XMH. Fuji added a 22-foot wing to the fuselage and enlarged horizontal stabilizers on the tailboom. On February 11, 1970, the XMH completed the first of fifty-seven successful test flights in various configurations. Although the program produced very promising data, lack of prospective military or civilian customers forced Fuji to cancel the project in 1973.

CONCLUSION

The success of helicopters in Vietnam, Africa, and the Middle East led world military leaders and manufacturers to seek even more advanced rotorcraft. Technological improvements in computers and composite materials presented the aerospace industry an opportunity to produce faster, more powerful rotorcraft with increased lethality and battlefield survivability. Night vision and navigation systems improved to the point that helicopters could operate in almost any weather conditions anywhere in the world. Many of the improvements were driven by military requirements, but helicopters intended for the commercial market reaped the technological benefits as well.

In addition to progress in what might be considered traditional helicopters, inventors continued to explore other designs of rotorcraft, including convertaplanes and tilt rotor machines. Although several engineers produced semisuccessful designs, only one aircraft, the Bell XV-15, produced enough interest to lead to a production aircraft. Even that aircraft, the V-22 Osprey, suffered the woes of most compound aircraft and did not reach full production during the period.

As in the past, helicopters and their crews continued to perform heroically in civilian life as well as in the military. In 1978 the award for heroism in aviation went to the crew of a British Airways Helicopters (BAH) S-61N. The crew braved the huge, windswept waves of the North Sea to rescue the entire crew of a sinking fishing trawler. The next year USAF 1LT. S. A. Stich and his ARRS crew, flying an HH-3, received the award for rescuing an Alaskan park ranger from a mountainside precipice. Combating low clouds, downdrafts, and wind shears, the crew completed an especially hazardous night rescue by hovering near the rocky mountainside until they managed to pluck the ranger from his perilous situation.

On October 25, 1983, Commander Harvey G. Fielding, flying a Sikorsky SH-3H on an SAR mission as part of Operation Urgent Fury, received an urgent call to rescue eleven wounded soldiers from the U.S. Army UH-60 downed by enemy fire. Fielding flew to the site, assessed the situation, and expertly landed his highly visible aircraft in a small cove. In only 3 feet of water, Fielding water taxied through enemy fire under the boughs of overhanging trees to rescue the crew and passengers of the Blackhawk. Fielding held his aircraft in place until his crew had loaded all the wounded aboard, then taxied clear and lifted the overloaded SH-3 off to transport the critically wounded to medical facilities. For his actions, the Navy awarded Fielding the Navy Cross ("The Wing Slip" 1984).

REFERENCES

Adcock, Al. *H-3 Sea King in Action.* Carrollton, TX: Squadron/Signal Publications, 1995.

Aircraft of the World: The Complete Guide. N.p./U.S.A: International Masters Publishers AB, licensed to IMP, 1996.

Anderton, David, and Jay Miller. *Boeing Helicopter: The CH-47.* Arlington, TX: Aerofax, 1989.

Apostolo, Giorgio. *The Illustrated Encyclopedia of Helicopters.* New York: Bonanza Books, 1984.

Bell Helicopter-Textron, http://www.bellhelicopter.textron.com. Accessed June 2002.

Boeing Company, http://www.boeing.com. Accessed January 2003.

Brehm, Jack, and Pete Nelson. *That Others May Live: The True Story of a PJ, a Member of America's Most Daring Rescue Force.* New York: Crown Publishers, 2000.

British Army Air Corps Association, http://www.aacn.org.uk. Accessed May 2002.

British Army Air Corps Historical Flight, http://www.rdg.ac.uk. Accessed May 2002.

British Army Air Corps Museum, http://www.flying-museum.org.uk. Accessed May 2002.

British Army Official Site, http://www.army.mod.uk. Accessed May 2002.

British Helicopter Museum, http://www.hmfriends.org.uk. Accessed May 2002.

Brown, David A. *The Bell Helicopter Textron Story: Changing the Way the World Flies.* Arlington, TX: Aerofax, 1995.

Carlson, Ted. "Marine Twin Hueys." *World Airpower Journal* 42 (autumn/fall 2000), pp. 134–143.

Chant, Christopher. *Fighting Helicopters of the 20th Century.* Christchurch, Dorset, England: Graham Beehag Books, 1996.

Cowin, Hugh W. *Military Helicopters.* New York: Gallery Books, imprint of W. H. Smith Publishers, 1984.

Donald, David, ed. *The Complete Encyclopedia of World Aircraft.* New York: Barnes and Noble Books, 1997.

Endres, Gunter, and Michael J. Gething, comps. and eds. *Jane's Aircraft Recognition Guide.* 3d ed. New York: HarperCollins Publishers, 2002.

Everett-Heath, John. *Soviet Helicopters.* London: Jane's Publishing Company, 1983.

Fort Rucker, AL. Home of U.S. Army Aviation, http://www.-rucker.army.mil. Accessed April 2002.

Fredriksen, John C. *Warbirds: An Illustrated Guide to U.S. Military Aircraft, 1915–2000.* Santa Barbara, CA: ABC-CLIO, 1999.

Gordon, Yefim, and Dimitriy Komissarov. "Mil Mi–24 'Hind.'" *World Airpower Journal* 37 (summer 1999), pp. 42–89.

Gunston, Bill, ed. *The Encyclopedia of Modern Warplanes.* New York: Barnes and Noble Books, 1995.

Gunston, Bill, and Mike Spick. *Modern Fighting Helicopters.* London: Crescent Books, 1996.

Harding, Stephen. *U.S. Army Aircraft since 1947.* Stillwater, MN: Specialty Press, 1990.

Heatley, Michael. *The Illustrated History of Helicopters.* New York: Bison Books, 1985.

Helicopter World, http://www.helicopter.virtualave.net. Accessed April 2002.

Helicopter's History Site, http://www.helis.com. Accessed June 2002.

"Helicopters of the U.S. Army," http://www.geocities.com/capecanaveral/hangar/3393/army.html. Accessed August 2002.

Higham, Robin, John T. Greenwood, and Von Hardesty. *Russian Aviation and Air Power in the Twentieth Century.* London: Frank Cass, 1998.

Hirschberg, Michael, and David K. Daley. *U.S. and Russian Helicopter Development in the 20th Century.* N.p.: American Helicopter Society International, 2000.

Igor Sikorsky Historical Archives, Inc., http://www.iconn.net/igor/index lnk.html. Accessed October 2002.

International Helicopters, http://www.globalsecurity.org. Accessed August 2002.

Junger, Sebastian. *The Perfect Storm.* New York: Harperperennial, 1999.

Kelly, Orr. *From a Dark Sky: The Story of the U.S. Air Force Special Operations.* Novato, CA: Presidio Press, 1996.

Keogan, Joseph. *The Igor I. Sikorsky Aircraft Legacy: The Chronology of Fixed-Winged and Rotary-Wing Aircraft of Igor I. Sikorsky and the Sikorsky Aircraft Company.* Stratford, CT: Igor I. Sikorsky Historical Archives, 2003.

Kuznetsov, G. I. *Kamov OKB 50 Years, 1948–1998.* Edinburgh, Scotland: Polygon Publishing House, Birlinn Publishing, 1999.

Lightbody, Andy, and Joe Poyer. *The Illustrated History of Helicopters.* Lincolnwood, IL: Publications International, 1990.

McGuire, Francis G. *Helicopters 1948–1998: A Contemporary History.* Alexandria, VA: Helicopter Association International, 1998.

Pearcy, Arthur. *U.S. Coast Guard Aircraft since 1916.* Shrewsbury, England: Airlife Publications, 1991.

Pember, Harry. *Seventy-five Years of Aviation Firsts.* Stratford, CT: Sikorsky Historical Archives, 1998.

Ripley, Tim. *Jane's Pocket Guide: Modern Military Helicopters.* London: Jane's, 1997.

Rogers, Mike. *VTOL Military Research Aircraft.* Somerset, England: Haynes & Co., 1989.

Royal Air Force Official Site, http://www.raf.mod.uk. Accessed May 2002.

Royal Navy Official Site, http://www.royal-navy.mod.uk. Accessed May 2002.

Russian Aviation Museum, http://www.ctrl-c.liu.se/misc/ram. Accessed November 2002.

Sikorsky Aircraft Corporation, http://www.sikorsky.com. Accessed June 2002.

Simpson, R. W. *Airlife's Helicopters & Rotorcraft: A Directory of World Manufacturers and Their Aircraft.* Shrewsbury, England: Airlife Publishing Ltd., 1998.

Soviet/Russian Helicopters, http://www.royfc.com/links/acft_coll. Accessed November 2002.

Spenser, Jay P. *Vertical Challenge: The Hiller Aircraft Story.* Seattle: University of Washington Press, 1992.

Stapfer, Hans-Heiri. *Mi-24 Hind in Action.* Carrollton, TX: Squadron/Signal Publications, 1988.

Swanborough, Gordon, and Peter M. Bowers. *United States Navy Aircraft since 1911.* London: Putnam, 1990.

Unusual Aircraft, http://www.unrealaircraft.com. Accessed March 2003.

U.S. Air Force Historical Research Agency, http://www.au.af.mil/au/afhra/. Accessed December 2002.

U.S. Air Force Museum, http://www.asc.wpafb.af.mil/museum. Accessed December 2002.

U.S. Army Aviation and Missile Command (AMCOM), http://www.redstone.army.mil/history/aviation/. Accessed February 2002.

Uttley, Matthew R. *Westland and the British Helicopter Industry, 1945–1960: Licensed Production vs. Indigenous Innovation.* London: Taylor and Francis, 2001.

"The Wing Slip." *Aviator's Post* #743. American Legion, Massapequa, NY. #576, May 18, 1984.

Wood, Derrick. *Jane's World Aircraft Recognition Handbook.* 4th ed. Colsdon, Surrey, England: Jane's Information Group, 1989.

Young, Warren R., et al. *The Epic of Flight: The Helicopters.* Alexandria, VA: Time-Life Books, 1982.

CHAPTER 6

Desert Shield/Storm and Beyond, 1990–2003

AFTER THE DISSOLUTION of the Soviet Union and the fall of the Berlin Wall, the end of the Cold War dramatically reduced funding for military helicopters, simultaneously increasing focus on generally smaller, civilian models. Of course not all funding ceased for military equipment, but worldwide, countries began to look to upgrade existing helicopters instead of buying new aircraft. In Russia military spending stagnated almost completely, causing manufacturers to convert military designs to civilian roles and seek international partners in order to survive in the world's marketplace. India, Japan, and the People's Republic of China began to produce more of their own aircraft, largely manufacturing Western designs as a licensee but also developing rotorcraft designed by their own engineers. European manufacturers joined together or merged with international companies to guarantee funding for new helicopters that met the military requirements of several countries. In the United States manufacturers also merged to modify and upgrade proven aircraft, as well as to produce a few new designs. Most politicians expected a "peace dividend" from military spending that would allow them to spend more on domestic issues, not realizing that the face of warfare was changing from massed armor engagements to smaller conflicts, requiring an adaptation of tactics and equipment. Because of recent vast changes in politico-military alignments and the international amalgamation of aerospace corporations, this last section is organized by primary aircraft function, not by nation or military affiliation.

COMBAT SUPPORT— PASSENGER HELICOPTERS

In 1988 two Kamov Ka-29RLD test aircraft, now known as the Ka-31, appeared on the Soviet carrier *Admiral Kuznetsov,* and publicly in 1995 at the Mosaero Air Show. The Ka-31, obviously a radar picket, or AWACS helicopter, displayed a large rotating E-801E Oko airborne EW radar mounted beneath the fuselage. The radar antenna folded flat against the fuselage for transit flight and landing. When in operation the antenna extended downward and rotated 360 degrees. The landing gear retracted in order not to interfere with the radar. The radar was known to track at least twenty independent air and surface threats. Some sources stated that the radar was capable of detecting as many as 200 targets and tracking 20 threats simultaneously, at ranges of more than 80 nautical miles for aircraft and 135 nautical miles for ships. The radar system automatically transmitted the targeting information over a secure data link to either a shipboard- or land-based command center. The Ka-31 also carried jammers to degrade or disrupt enemy communications and acquisition radars. Both the Russian and Indian navies (as of 2001) operated versions of the Ka-31. In the early twenty-first century Kamov began to promote an enhanced tactical version of the Ka-31, which, reportedly, detected low-flying aircraft, helicopters, and cruise missiles. The system, used by Army or Air Force command centers, automatically transmitted the collected reconnaissance information to C&C centers, Unmanned Aerial Vehicles (UAV), jet fighters, or attack helicopters for immediate attack.

In the late 1990s the Kamov Bureau introduced the Ka-226, another modernized commercial helicopter in the lineage of the naval Ka-26 Hoodlum. The aircraft retained the usual three-bladed coaxial rotor system, but with blades manufactured from polymeric composite materials, semirigidly attached to the hub with a bearingless torsion bar, as on the Ka-29/31. Two Allison 250-C20R 450-horsepower turboshafts, mounted above the nine-seat passenger cabin, turned the 42-foot, 8-inch counter-rotating rotors. The Ka-226, de-

signed mainly for passenger/cargo transportation and EMS operations, included options such as an all-glass cockpit and Western
avionics, which met all governmental requirements for certification
as a commercial helicopter, including the U.S. Federal Aviation Regulations (FARs). The Ka-226 cargo version's capabilities included
hauling 3,000 pounds of internal cargo, or a 3,300-pound slingload.
Brochures listed maximum speed as 110 knots, max range as 325
nautical miles, and a ceiling of 20,000 feet. Russia and several former Soviet satellite countries operate the Ka-226, with Kamov in an
active quest for international customers.

In 1999, Kamov, in association with Pratt & Whitney of Canada,
introduced the Ka-115. The light multipurpose helicopter incorporated the traditional Kamov design but sat on a skid landing gear instead of wheels. The five-door cabin held up to five passengers or a
2,000-pound payload. A single GTD PW/K 206D 550-horsepower
turboshaft drove the 33-foot, 2-inch rotor system and the aircraft to
a maximum speed of 135 knots and a range of 420 nautical miles.
Kamov designed the single pilot machine to operate in all climatic
conditions, with deicing systems for both the engine inlets and rotor
blades, and a heating/air conditioning system to guarantee passenger comfort. The Ka-115's communications and navigation equipment met all international standards and was marketed for a variety
of roles, including passenger and light cargo, EMS, SAR, police patrol, and charters.

In 1993, Kamov initiated a program to produce a UAV, previously
known as drones and Remotely Piloted Vehicles (RPVs), the Ka-37
remotely piloted helicopter. Designed for battlefield reconnaissance
with TV and IR sensors, radio intercept and relay, ecological surveys,
and delivery of foodstuffs or medical supplies to remote or dangerous locations, the Ka-37 retained the counter-rotating coaxial rotor
system favored by the Kamov Bureau. The bureau equipped the pilotless vehicle with computer technology and modular components
that allowed the operator to preprogram flight paths and mission requirements, along with the capability for the operator to fly the helicopter from a digitized ground control station if necessary. The aircraft, ground control station, and power generators were designed to
fit into a specialized transportation container. A 60-horsepower piston engine powered the little aircraft, which had a maximum payload of 110 pounds and a maximum endurance airspeed of 60 knots.
The Ka-37 reached altitudes of 8,000 feet, but line of sight and an
endurance of only about one hour limited radius of operations. The

bureau planned to introduce an improved Ka-137 multipurpose re-motely piloted helicopter with enhanced capabilities in the near future.

Bell Helicopter's esteemed Huey retained a reputation for relia-bility and service for thirty years, with several countries, including the United States, employing the helicopter into the twenty-first century. In Indonesia, IPTN built the Model 412 under license as the Nbell-412. In Japan, Fuji Industries developed a unique Japa-nese variant of the UH-1H, appropriately designated UH-1J. A sin-gle Allison T53-L-703 1,800-horsepower turboshaft powered the UH-1J, which also incorporated the composite swept-tip rotor blades of the Bell Model 212. Fuji included a vibration reduction system, infrared countermeasures, and an NVG-compatible cockpit in the updated Huey. In 1993 the Japanese Self-Defense Forces ac-cepted the first of seventy-eight UH-1Js, with deliveries completed in 1998. Fuji also manufactured ninety-four versions of the AH-1F Modernized Cobra, powered by the Kawasaki Lycoming T53-L-703 engine.

During Operation Desert Shield/Storm (1990–1991) the U.S. Army deployed 115 OH-58d aircraft to Southwest Asia. The heli-copters flew almost 9,000 hours in combat action and posted a full mission capability rate of 92 percent, better than any other scout/attack aircraft in the theater. Into the twenty-first century the OH-58D enjoyed the lowest ratio of maintenance hours to flight hours of any Army combat aircraft. OH-58Ds accompanied armored cav-alry units, which had only OH-58Cs assigned, to provide enhanced battlefield surveillance and target acquisition for the AH-1 Cobras. The "D" models also provided protection from enemy aircraft with their Stinger AAMs. In addition to Desert Shield/Storm, OH-58Ds participated in operations in Haiti, Somalia, and Bosnia during the 1990s.

In the late 1990s the U.S. Army converted all OH-58D aircraft to the armed Kiowa Warrior. Along with armed reconnaissance, com-mand and control, and target acquisition/designation under adverse conditions, the Kiowa Warrior, equipped with the Multipurpose Light Helicopter (MPLH) kits, was capable of multimission flexibil-ity such as limited troop transportation, emergency medical evacua-tion, and small slingload operations. During the same period Kiowa Warriors began to supplant both AH-1 Cobras and OH-58Cs in U.S. Army Air Cavalry and light attack helicopter units.

In Operation Iraqi Freedom (OIF) in 2003, OH-58D Kiowa War-riors provided Close Combat Attacks (CCA) in support of U.S.

troops in the battles of An Najaf, Karbala, Al Hillah, southern Bagh-dad, and Mosul. Aviation commanders selected the Kiowa Warrior as the best aircraft to support infantry in urban combat situations. Not as heavily armed as the AH-64, the OH-58D, with its doors re-moved, provided the crew with superior visibility to identify and de-stroy individuals skulking among the alleyways and buildings of the cities. The helicopter's low profile and maneuverability increased its combat survivability in an urban environment. Usually employed in pairs, the Kiowa Warriors provided both fire support and a forward screen for troops advancing along city streets. OH-58D pilots used a combination of running and diving fire to employ their .50-caliber machine guns and rockets most effectively. The Kiowa Warriors ac-cumulated well over 5,000 hours of combat flying and received credit for more than 200 enemy vehicles, as well as many artillery and air defense positions. Other than UAVs, the OH-58D provided the most effective reconnaissance of urban areas during the major combat stage of OIF, identifying concealed enemy positions and nu-merous weapons caches. Several OH-58Ds sustained damage from enemy fire and one was shot down, but most were back in service in a matter of hours.

The Kiowa Warrior, however, suffered from several shortcomings that affected mission accomplishment and safety. Although the MMS provided almost unparalleled penetration of darkness and ad-verse weather, and with the doors removed the crew had good visi-bility around the aircraft, the pilots, nonetheless, flew with NVGs, which limited visual acuity and peripheral vision. As a result, night-time accidents occurred at a higher rate than in the AH-64 Apache equipped with the PNVS. The added weight of armament dimin-ished the OH-58D Kiowa Warrior's autorotative capability, identi-fied as marginal by Army test pilots during the AHIP development, and required extraordinary pilot expertise to complete a safe landing in the event of an emergency. The Army alleviated this problem somewhat by adding supplemental crew restraints, air bags, and crashworthy seats (requested in AHIP development by Army test pi-lots but not purchased initially by the DoD).

Until the early 1990s the Taiwan military made use of OH-6As as observation helicopters. In 1991, under the program code-named *Lu Peng*, Taiwan requested to purchase twelve OH-58Ds, along with eighteen AH-1Ws, from the United States to replace the OH-6A. The Taiwan government increased the order to twenty-six OH-58Ds, enough to equip two armed scout units. On June 24, 1998, AIDC in Taiwan contracted to assemble another thirteen OH-58Ds, ordered the previous year. In November 1999, AIDC completed the first of these aircraft, which were shipped to Bell for test flight. The Taiwanese OH-58Ds resembled the U.S. Kiowa Warrior, armed with .50-caliber machine guns, Hydra 70 2.75-inch FFARs, four Hellfire ASMs, and the ALQ-144 IR jammer. All helicopters were delivered by the end of 2001.

In the last decade of the twentieth century Aerospatiale executives took advantage of several business opportunities to increase their market share of the world's helicopter market. In 1990, Aerospatiale redesignated all its military helicopters with a "5"—for example, the Super Pumas became the AS-532 Cougar, the "5" clearly distinguishing the military variants from the civilian. On January 31, 1992, Aerospatiale joined with MBB to create Eurocopter Holdings. The new corporation included Daimler-Benz-Chrysler Aerospace, which controlled 40 percent of the company, Aerospatiale holding the remaining 60. The management company, Eurocopter SA, fell under Eurocopter Holdings (75 percent) and Aerospatiale (25 percent). The same year Aerospatiale introduced the AS-532 Cougar Mark II, the largest and most powerful of the successful Puma/Cougar line. In addition to a streamlined fuselage, capable of accommodating twenty-nine combat loaded troops and a crew of four, an improved "Spheriflex" rotor system and two Turbomeca Makila 1A2 2,100-horsepower turboshafts provided the Cougar with more speed, maneuverability, and lift capacity than any previous model. The Cougar came equipped with a fully NVG-compatible cockpit, a four-axis autopilot, and GPS navigation system; it was fully prepared for all weather operations, including electric de-icers for the main and tailrotors. To enhance the Cougar's battlefield survivability, the French company installed IR diffusers on the engine exhausts and equipped the helicopter with an IR jammer, a radar warning system, and flare and chaff dispensers. In addition to utility transport, ASW, and SAR, the Cougar appeared in VIP, Electronic Warfare (EW), and surveillance configurations. Equipped with the Thomson-CSF target system, the Cougar flew behind friendly lines and scanned enemy territory with an X-band radar. The radar located targets up to 200 km behind enemy lines within an accuracy of 40 meters. The radar contained a Moving Target Indicator (MTI) that identified vehicles, boats, or helicopters moving as slowly as 7 kph. The aircraft system automatically transmitted the intelligence to a ground station that relayed it to the battlefield commander. The prototype system received actual combat testing during Desert Storm and in Kosovo, proving its interoperability with NATO and the U.S. Joint Surveillance Tactical Aperture Radar (JSTAR). The French Army received the last of four Cougars with this system in 1997.

During the mid-1980s, MBB began design work on a helicopter to replace the Bo 105 and, in the autumn of 1988, flew the prototype Bo 108. After the amalgamation of Eurocopter, the aircraft became the testbed for the EC-135. From its silhouette the new EC-135, which first flew in early 1994, appeared to be the front seven-passenger fuselage of the Bo 108 with a tailboom and ten-blade Fenestron borrowed from Aerospatiale. Eurocopter used the most advanced composite technologies in constructing the helicopter, installing a 33-foot, 6-inch "Spheriflex" rotor system above an 11-foot, 6-inch cabin, which incorporated two litters and two seats for medical attendants in the EMS role, or five passengers in the luxurious VIP suite. Powered by two 600-horsepower Turbomeca Arrius 2B engines, the EC-135 demonstrated a maximum speed of 140 knots, a range of 330 nautical miles, 430 with long-range tanks, and a service ceiling of 10,000 feet. Customers had the option of a conventional or the Avionique Nouvelle all glass cockpit. The helicopter received its certificate from the German government in 1996, with initial customer deliveries the next month. Eurocopter America also manufactured the EC-135P2 with two Pratt & Whitney PW.206B 620-horsepower turboshafts for certain customers, especially those in North America. By the end of 2003, twenty-five countries operated more than 260 models of the EC-135, including armed forces, police departments, various corporations, and EMS providers. The military variants of the EC-135 included composite armor and other military equipment, including a cargo hook capable of lifting a 2,750-pound slingload.

The renamed operating companies of the Eurocopter conglomerate, France and Eurocopter Deutschland, inherited, in addition to existing production aircraft, several military development projects, including the NH 90 and PAH-2 Tiger. The NH 90 began under governmental agreements that joined Aerospatiale, MBB, Agusta, and Fokker in a new joint venture company called NH Industries. In September 1992, Nahema, the NATO agency representing France, Italy, Germany, and The Netherlands, opened a feasibility study to design and develop a medium lift transport helicopter capable of lifting twenty combat-loaded troops, and the flexibility to fulfill naval ASW and SAR roles. NH Industries expected orders for more than 700 of the two major variants of the proposed NH 90, a Tactical Transport Helicopter (TTH) and NATO Frigate Helicopter

(NFH). Designers of the NH 90 incorporated composite materials in "stealth technology," modularity of electronic components, full-color Multi Function Displays (MFD) in an "all glass" cockpit, and dual integration of the 1553B digital data bus to produce a "fly by wire" helicopter with a low radar signature, capable of single-pilot operations in day/night and adverse weather conditions. The TTH easily handled twenty troops and their equipment or 5,500 pounds of internal cargo. As a medevac the NH 90 accommodated twelve litters and medical attendants with their associated equipment. Additionally, the helicopter could be quickly configured to perform SAR, VIP, EW, and special operations missions, including paradrops. The NFH version included a ship landing system, search radars, a dipping sonar, and weaponry for autonomist ASW and anti-ship warfare, but could also be used for SAR, vertical replenishment, and troop transport.

In mid-1995, Eurocopter France completed the first of five flying prototypes, with the first taking flight on December 18. The NH 90 featured a titanium main rotor hub with elastomeric bearings and an advanced composite 53-foot, 6-inch four-bladed main rotor. The 52-foot, 10-inch fuselage housed two 1,830-horsepower turboshafts, MTU/Rolls-Royce/Turbomeca MTR322 (TTH) or General Electric T700-T6E (NFH), a transmission with "thirty minute dry run" capability, and retractable landing gear. Designers expected the helicopter to have a maximum speed of 170 knots, and a range of more than 500 nautical miles.

On June 8, 2000, NH Industries received official notification to place the NH 90 into production, and the next day it announced orders for 243 of the helicopters. Between January and November 2001, Portugal, Finland, Norway, and Sweden also placed orders for the NH 90, with expected deliveries in 2003.

During the 1990s the Westland Lynx continued to reliably serve U.K. forces. In Desert Storm, RN Lynxes scored fifteen confirmed hits against Iraqi vessels with BEA Sea Skua missiles, destroying most of Iraq's patrol boats. British Army and Navy Lynx crews flew the equivalent of seven months of combat action in only six weeks, with the Lynx performing superbly in the desert sands. In Bosnia, British crews flew numerous combat patrols and lifesaving missions, including delivering vital medical supplies to wretched, embattled refugee camps.

In April 1994, Westland became Westland Plc and part of the GKN Group, which holds manufacturing interests in forty countries. In 1995, Westland delivered the new Super Sea Lynx HMA

Mk 8 (export version Mk 88A) to the RN and an upgraded battle-field Lynx Mk 7 to the British Army. The Super Lynx included a 42-foot BERP rotor system, which resulted in a faster, more agile helicopter, capable of carrying heavier loads, including a 3,000-pound slingload, to ranges of 285 nautical miles. Westland also installed an entirely new computer-integrated tactical combat cockpit and advanced avionics in both versions. The instrumentation included MFDs for flight, navigation, and tactical displays, linked to a flight director, and GPS navigation system. The naval version included a new chin-mounted Sea Spray radar capable of a 180-degree scan. Upgraded armament included either Marte Mk 1 or Penguin AAMs. More powerful Rolls-Royce GEM 42-11 1,100-horsepower engines, with an improved passive IR suppression system, powered the new aircraft to 138 knots. Argentina, Brazil, Germany, Denmark, France, Malaysia, The Netherlands, Nigeria, Norway, Portugal, South Africa, Oman, South Korea, Thailand, and the United Kingdom operate more than 400 versions of the Lynx.

As early as the late 1970s the British Royal Navy began a search to replace its technologically aging Sea King ASW helicopters. In response GKN Westland initiated the WG 34 project. Simultaneously, the Italian Navy considered replacing its own Sea Kings. Consequently the two governments agreed to develop a new ASW helicopter jointly. The agreement led to the establishment of European Helicopter Industries (EHI) in June 1980 (AgustaWestland on February 12, 2001) and a joint funding arrangement on January 25, 1984. Westland and Agusta spent almost eight years in an assiduous design program before the first EH 101 prototype took to the air on October 9, 1987.

The medium lift multirole helicopter, capable of carrying up to fifty-five survivors in the SAR variant, depended on three General Electric CT7-6 2,000-horsepower turbines, mounted atop the fuselage, to turn the advanced 61-foot, five-bladed main and four-bladed tailrotors, fabricated with a carbon-glass skin and a nomex-foam honeycomb core. The EH 101 cruised comfortably at 150 knots and exhibited a range of 460 nautical miles without the available auxiliary fuel tank; designers fitted the military variants with a hover in flight refueling (HIFR) system. The external hook safely lifted loads up to 12,000 pounds. Sponsons along the lower aluminum-lithium fuselage housed a retractable tricycle landing gear. EHI's program called for nine preproduction aircraft, two utilized for civil certification.

The cockpit contained dual controls and the latest in advanced instrumentation. Six full-color MFDs linked to an advanced flight management system, including an Automatic Heading-Reference System, AHRS, weather radar, and integrated avionics suite provided necessary information and control for certification as a single-pilot IFR helicopter. Deicing equipment on engine inlets and rotor blades added an all-weather component to the helicopter's capabilities.

The long development cycle resulted from the myriad of requirements stipulated by the two militaries. The RN required a multipurpose ASW model, the RAF an SAR/utility aircraft, while the Italian Navy demanded three variants, an ASW version, a surveillance radar picket aircraft, and a naval utility transport with both folding blades and tail section, along with a rear-loading cargo ramp. With folding blades and tail, the EH 101 could operate from small frigates and fit into the same underdeck hangars as the Sea King. In March 1996 the first of forty-four EH 101 Merlin HMA1s ordered by the Royal Navy took flight. On May 27, 1997, the first of these ASW helicopters arrived at an RN squadron. Shortly afterward, the RAF received the first of twenty HC3 utility helicopters. Canada ordered fifteen EH 101 Cormorant SAR helicopters to replace the Canadian Armed Forces' fleet of Boeing CH-113s. Other international orders included the following: Denmark, fourteen search and rescue and troop transport variants; Portugal, twelve SAR and fishery protection aircraft; and Japan, fourteen for utility transport and mine countermeasures.

Shortly after the Cormorant went into service with the Canadian military in 2002 it performed its first rescues. On July 28 a crewman aboard the container ship *Cynthia Melody* received a serious head injury, and the ship's captain called for immediate medical assistance. A Cormorant, piloted by Captain Jennifer Weissenborn, flew through "poor weather" to rendezvous with the ship about 115 miles north of Vancouver Island. Hovering over the deck, the crew lowered two search and rescue technicians onto the ship to stabilize the patient and then hoisted him aboard the helicopter by a Stokes litter. The Cormorant crew then delivered the injured man to a hospital in Comox, Nova Scotia. The second rescue occurred two days later, when a Cormorant landed in inhospitable terrain to rescue a father and his two sons after they were forced to put down their chartered plane on account of bad weather.

The commercial version of the EH 101 flew in 1997 and was offered in two variants, one with a rear-loading cargo ramp and the Heliliner version, with plush seating for thirty passengers. The cabin featured full standing room, a wide central aisle, overhead storage and a large baggage compartment, as well as a powerful cooling/

heating system for passenger comfort. In the medical evacuation role the cabin held sixteen stretchers, along with medical equipment and attendants. Lockheed Martin, leading "Team U.S. 101," proposed the VIP version as the Marine One presidential transport helicopter, but no decision was forthcoming at the time of publication. In addition to producing the EH 101, Westland entered into developing a variant of the Sikorsky S-70 Blackhawk powered by RTM-322 engines, intended for the Saudi *Al Yamamah II* program.

In October 1989 the Sikorsky UH-60L began to replace the UH-60A as the standard production utility helicopter for the U.S. Army. Experience around the world convinced the Army that the Blackhawk required significant increases in performance to meet the needs of soldiers in combat. The L model incorporated uprated T700-GE-701C engines that included new IR shrouds and digital electronic fuel controls that increased available power from 1,560 horsepower to 1,940 horsepower per engine; more durable gearboxes; increased pitch for the tailrotor to allow pilots to make use of the increased engine power; SH-60B flight controls; and the External Stores Support System (ESSS), a removable wing pylon with four stations for external fuel tanks or other stores, including sixteen Hellfire ATGMs. The ESSS fuel system extended the range of the UH-60L to 1,140 nautical miles. The ALQ-144 IR jammer and M-130 chaff/flare dispenser came standard on the new Blackhawk. The improvements increased cruise airspeed to 152 knots, improved high-altitude and hot weather operations, and increased the maximum gross weight from 20,250 to 22,000 pounds. The UH-60L also gained the ability to sling a 9,000-pound load High Mobility Multiwheeled Vehicle (HMMV), which replaced the jeep as the Army's primary frontline vehicle.

After the October 1993 command and control debacle in Somalia, the U.S. military realized that Special Operations Command (SOCOM) required specialized helicopters with sophisticated communications and navigation systems, as well as heavier firepower. Despite the heroic efforts of the OH-6 "Little Bird" pilots of the 160th Aviation Regiment, who poured minigun fire into the rebel fighters all night, their aircraft lacked the punch to drive the attackers away from the besieged Army Delta Force and Rangers. Although ten fully armed AH-1F Cobras sat on standby a few minutes away, arrogant special operations officers refused to permit the heavily armed Cobras to enter the desperate struggle, resulting in needless deaths of U.S. soldiers. As a result, the DoD ordered the Sikorsky MH-60G PaveHawk, a USAF special mission and deep penetration rescue helicopter with in-flight refueling probe, special mission avionics, including Side Looking Infrared (SLIR), GPS, and a computerized flight direction system linked to an integrated multifunctional all glass cockpit. The Army received the MH-60K, similar to the MH-60G with a chin-mounted FLIR replacing the SLIR, and external fuel tanks, a 30-mm cannon, and Stinger missiles installed on the ESSS. The MH-60K mounted .50-caliber doorguns, and some carried two 20-mm cannons. Uprated T-700-GE-701C engines increased maximum gross weights and hot weather performance.

During Desert Shield/Storm, Blackhawks flew thousands of combat hours and performed superbly, with only one lost to enemy fire. The UH-60 again proved its rugged dependability in the first Persian Gulf War, but when the United States sent forces into Afghanistan in 2002 the mountainous terrain severely limited the Blackhawk's efficiency. Like many Soviet helicopters in the 1980s, the UH-60 proved underpowered to operate at high gross weights in the 18,000-foot mountains, and a loss of tailrotor authority caused a loss of directional control at critical points in the flight profile. The Army required the UH-60 to lift and move 10,000 pounds at 4,000 feet at 95 degrees Fahrenheit, but higher-density altitudes and temperatures reduced the useful load that the helicopter could carry. Instead of its fleet of UH-60s the Army used the more powerful CH-47 Chinooks to conduct combat assaults, as well as ferry troops and supplies.

To remedy the problem the Army plans to replace all T700-GE-700 (1,622 horsepower) on the UH-60A and the T700-GE-701C (1,890 horsepower) on the UH-60L with the T700-GE-701D (2,000 horsepower) on the UH-60M, expected to be delivered in 2005. The UH-60M will also have an uprated transmission and gearboxes to utilize the increased engine power. In addition to the United States, Argentina, Australia (assembled by Hawker de Havilland), Brunei, China (Taiwan), Colombia, Japan (built by Mitsubishi Industries), Egypt, Greece, Hong Kong, Iran, Jordan, Mexico, Morocco, Philippines, Saudi Arabia, South Korea, Thailand, Turkey (fifty assembled by Havacilik ve Uzay Sanayi A.S. in Turkey), and the United Kingdom used versions of the 2,433 S-70/UH-60s produced by 2003.

In 1992, Sikorsky introduced a new helicopter but delayed the prototypes until 1998. Sikorsky Aircraft joined a consortium that included Mitsubishi Heavy Industries, Gamesa (Spain), AIDC (Taiwan), and Jingdezhen Helicopter Group (China) to produce the proposed commercial and military aircraft, which eventually won the

coveted Collier Trophy for excellence in aviation design in 2002.
Based on the successful S-70 airframe, the S-92 Helibus was intended to rival the Super Puma as a military transport and replace
the S-61N in oil exploration, SAR, and EMS applications. Two
newly developed General Electric CT7-8C 3,000-horsepower turboshafts turned a 56-foot, 4-inch S-70 four-bladed composite main
and four-bladed tailrotors, with anti- and deicing equipment. The S-
92 featured a modern computerized cockpit, intended to comply
with all certification requirements for single-pilot IFR operations. A
redesigned fuselage included a nineteen- to twenty-two-passenger
cabin and sponsons along the lower sides for a retractable landing
gear and long-range fuel tanks. Helibus trials indicated a cruise
speed of 155 knots. Sikorsky planned the first commercial Helibus
deliveries for 2002.

In 1986 the Mil design bureau first flew its light, aerobatic training helicopter. Two years later Mil pilots demonstrated the "T-tailed"
Mi-34 Hermit at Air/Space America. Initially intended for domestic
users, the rapid alteration of Russia's economic dynamics, and international interest in the 3,200-pound helicopter, convinced Mil to
produce the Mi-34S export version. Similar to the Aerospatiale
Ecureuil, the Helix seats four in its diminutive cockpit. Powered by
either a single Russian M-14B26B or Textron-Lycoming TIO-540
350-horsepower piston engine mounted behind the cabin, turning
the 32-foot, 10-inch four-bladed main rotor, the Mi-34S recorded a
maximum speed of 120 knots. Mil experimented with twin VAZ-430
rotary piston engines in aircraft for domestic use but considered the
Mi-34A, upgraded with an Allison 250-C20R turboshaft and heavier
transmission, the most lucrative design. Mil Light Helicopters began promoting this aircraft internationally in the late 1990s. In
1992, Mil signed an agreement with the Anglo-U.S. Brooke Group
to provide parts and worldwide technical services for Mil helicopters
through its subsidiary Mil-Brooke Helicopters, headquartered in
Miami, Florida.

In July 1994 the Hiller family and a group of Asian investors reac-
quired the Hiller name from Rogerson Aircraft Corporation and renamed the company Hiller Aviation Corporation. In September
1995 the company acquired the type certificates and conversion kits
for the Soloy turbine conversions of the UH-12 and Bell 47. In
1996 the new management returned the UH-12E to production
with thirty machines at Marina, California, for the Mexican Navy
and Summit Helicopters Inc.

COMBAT SERVICE SUPPORT HELICOPTERS

Coalition forces, led by the United States and the United Kingdom, deployed more than 160 Chinooks to the Persian Gulf for Operation Desert Shield and logged in excess of 16,000 hours of mission time during that conflict. In addition, the armed forces of Canada, Italy, South Korea, and Thailand ordered export versions of the BV-414, most of which received factory upgrading. Between September 1991 and April 1993, Boeing rebuilt eighteen Spanish Army CH-47Cs to meet CH-47D standards. The Netherlands bought seven Canadian Ch-47Cs, and Boeing upgraded them to CH-47s as well. Boeing Vertol also granted a license to manufacture the Chinook to the Italian associate company of Agusta, Elicotteri Meridionali SpA. The consortium produced thirty-four aircraft for the Italian Army and sold several Chinooks to Egypt, Greece, Iran, Libya, and Morocco. In Japan, Kawasaki received a license to build the CH-47J, also a standard CH-47D Chinook, powered by two Mitsubishi-Lycoming T55-K-712 turboshafts. Deliveries of the first of forty-five Chinooks to the JASDF and JGSDF began in 1986.

In the late 1980s, responding to a U.S. Army requirement for a medium lift helicopter for clandestine operations, Boeing produced special operations versions of the CH-47D, the MH-47D, and the follow-on MH-47E. The company revamped the machines with a Model 234 nose section containing a terrain-following radar, an integrated advanced avionics system, and air defense radar jammers. Increasing the self-defense capabilities of the Chinooks, Boeing installed window-mounted 7.62-mm miniguns and Stinger AAM missile racks. The company also added increased fuel capacity and booms for in-flight refueling. On June 1, 1990, SOCOM officially took delivery of the first MH-47D.

In 1998 a definitive new helicopter from the Kamov OKB in Russia made its maiden flight. The Ka-60 Kasatka (Killer Whale) differed radically from previous Kamov designs in that the helicopter had a "conventional" main and antitorque rotor system rather than the "standard" Kamov coaxial rotors. A four-bladed 44-foot, 3-inch composite main rotor and eleven-bladed "fenestron-type" tailrotor, manufactured from carbon-reinforced Kevlar, provided lift and directional control. The Ka-60 was designed as a medium lift battlefield helicopter, capable of delivering sixteen combat loaded troops or 6,000 pounds of supplies, either internally or by slingload. In a medical evacuation role the cargo compartment accommodated six stretchers and three medical attendants. Two Rybinsk RD-600V

1,280-horsepower turbines, manufactured by NPO Saturn, powered the helicopter. Kamov planned to offer an export Ka-60 with two General Electric CT7 engines.

Kamov constructed the Ka-60 with battlefield survivability as the foremost priority. With rotors and fuselage fabricated of more than 60 percent composite materials, the Kasatka had a low radar and IR signature, enhanced by special absorbent exterior coatings and IR suppressers on the engines. The swept-tipped rotor blades were designed to sustain multiple 23-mm hits without disintegrating, and the foam-filled fuel cells prevented explosions. The transmission reportedly resisted up to 12.7-mm rounds and had a limited "dry run" capability. Kamov included redundancy in all hydraulic and electronic systems. The two pilots sat in armored seats and had a Pastel radar warning receiver and an Otklik laser warning system; the Arbalet millimeter wave radar provided instant battlefield and weather information. Armament included two 7-round 80-mm rocket pods and either a 7.62-mm or 12.7-mm machine gun mounted at the rear of each cargo door for self-defense. The Kasatka rested on a rugged retractable tricycle landing gear, with the option of inflatable pontoons for emergency over-water operations; it had a top speed of 160 knots and a range of 325 nautical miles. The Russian military expected the first deliveries of the Kasatka sometime in 2004.

The Kamov Bureau developed the Ka-62 civil version of the Ka-60 using the same high-tech fuselage and rotor system. The Ka-62 incorporates the composite construction and the fuel-efficient RD-600 engines of the military version. The machine meets most foreign government requirements for both VFR and IFR operations, including blade deicing and engine compartment fire suppression. Capable of carrying sixteen passengers in the spacious air-conditioned cabin, the Ka-62 is also designed to fill the EMS role when a larger helicopter is required. The cargo version is equipped for slingload operations and has the wide side doors of the Ka-60. Upon customer request Kamov plans to furnish the Ka-62 with Western-built engines and avionics. Kamov hopes to find customers who require SAR, border patrol, or offshore aircraft.

Beginning in the late 1990s the Mil design bureau entered into several agreements with Western aerospace corporations, either to produce aircraft jointly or to acquire components and technology for its own designs. In 1997, Mil joined the Franco-German firm Eurocopter to produce the medium lift (30,000-pound) thirty-passenger Mi-38 to replace the aging Mi-17. Two TV7-117V turboshafts powered the six-bladed composite rotor system on the Mi-38. Other improvements included a computerized flight instrumentation and navigation system, a larger passenger cabin, and retractable tricycle landing gear. No production information was available in 2005.

Other joint projects in which Mil participated were the ten- to twelve-passenger Mi-54 in the 2,800- to 3,000-pound range, in both executive and utility versions. To replace the decrepit Mi-6 and Mi-10K, Mil proposed the heavy lift Mi-46, in both transport and flying crane configurations. In addition to its civil helicopter designs/proposals, Mil proposed a new military helicopter, the Mi-40, basically an enlarged Mi-28 with an enlarged cabin to accommodate eight troops; fuselage sponsons incorporate a new retractable landing gear and weapons systems. Western companies such as Eurocopter, Allison, Lycoming, and Pratt & Whitney agreed to provide turbine engines to Mil in order that the Russian helicopters will be accepted by European and North American customers.

ATTACK HELICOPTERS

In 1977 the United Nations placed sanctions against South Africa, and Atlas Aircraft became the only source for both maintenance and new aircraft for the South African National Defense Force. The company's gunship version of the Puma proved too large and not nearly agile enough for an attack helicopter. As a result Atlas developed a concept helicopter based on the Alouette III and designated the XH-1 Alpha. A pilot and gunner sat in a "stepped-up" tandem cockpit, from where the gunner, by using his helmet sight, controlled a 20-mm cannon mounted in a chin turret. A single Turbomeca Artouste IIIB 570-horsepower turboshaft powered the Alpha, which flew for the first time on February 27, 1986. Failing to meet expectations, the machine was retired to the SAAF Museum in Pretoria.

The XH-1 did, however, supply Atlas's aeronautical engineering department (by this time Denel) the requisite data to design the XH-2 Rooivalk, later designated the CSH-2 and then AH-2. On February 11, 1990, the first of two Rooivalk prototypes, a twin-engined tandem-seated gunship, made its maiden flight. Typical of most antitank designs, the pilot sat behind the copilot/gunner, both in armored seats. Two Denel-manufactured versions of the 2,000-horsepower Turbomeca Makila 1A1 engine, known as the Topaz, powered the 51-foot, 1-inch composite four-bladed main and five-bladed conventional tailrotors. State-of-the-art shrouds and suppressors masked the aircraft's IR signature. Aerospatiale—later Eurocopter—provided the same transmission and other dynamic components used on the Puma.

Denel equipped the production Rooivalk with the most advanced cockpit, with two MFDs and an additional navigation display. Targeted FLIR and TV sensors provided the crew with a night vision system. Armament included the F2 20-mm cannon in a chin turret, and 70-mm rocket pods, or ZT6 Mokopa ATGWs on the inner pylons of the AH-2's stub wings. Mistral AAM racks were installed on the outer pylons. Both the domestic and export versions of the aircraft flew at 167 knots and had a range of 720 nautical miles. Denel delivered the first of a dozen Rooivalks ordered by the SAAF in 1999.

In the late 1980s the Soviet Union, and then Russia, began to upgrade the Mi-24 Hind D to the Hind E and F models. The airframe and rotor remained basically the same, but with some modifications, especially in equipment. A titanium rotor hub and lightweight composite rotor blades replaced the outdated steel components, and self-sealing, foam-filled fuel tanks replaced the aluminum tanks. Modifications included the capacity to carry drop tanks on the wing pylons. Most Hinds retained the 12.7-mm Gatling gun in the turret, but some Hind Es carried a twin-barreled 23-mm cannon in the turret. The new AT-6 Spiral ATGM replaced the AT-2. Armament also included two GSh-23L twin 23-mm cannon pods on the wings. The Hind F, believed to be designed as an air-to-air helicopter for shooting down opposing attack helicopters, carried a fixed 30-mm cannon on the right side of the fuselage, some with the turret and gun replaced by a search radar.

The Mi-24D was generally restricted to daytime operations, but the Hind enhancements included equipment and capabilities that permitted night and all-weather operations. Mil engineers installed a HUD, FLIR, low-light TV, laser rangefinders/designators, and missile guidance system to support the AT-6 missiles. The Hind E/F, and export version Mi-35, included new avionics systems, weather radar, auto pilot, GPS, all within an NVG-compatible cockpit. IR suppressers shrouded the engine exhausts, and radar warning receivers alerted the pilots to activate new flare and chaff dispensers. Both the main and tailrotors were electrically deiced. Modifications to the Mi-35M, a radically modified export version of the Hind, included two TV3-117YMA 2,250-horsepower turboshafts, Mi-28 rotor blades and hub, and a fixed landing gear. Ataka and IGLA-V missiles replaced older ATGMs as the main armament.

Other variants of the Hind included a training aircraft with dual controls; a reconnaissance aircraft with cameras and video downlink; an EW version with radio/radar jammers and radio intercept/location equipment; and a specialized aircraft for monitoring chemical, biological, and nuclear environments, with an overpressure system to protect the crew from contaminants. At one time at least thirty-four countries operated variants of the Hind, including Armenia, Algeria, Angola, Belarus, Bulgaria, Cambodia, the Commonwealth of Independent States, Cuba, Czechoslovakia, Czech Republic, Ethiopia, Georgia, Federal Republic of Germany (East Germany), Hungary, Iran, Iraq, Libya, Mongolia, Mozambique, Nicaragua, North Korea, Peru, Poland, Slovakia, South Yemen, Syria, Ukraine, Vietnam, and Zimbabwe.

Both the U.S. Army and USMC AH-1 Cobras participated in Operation Desert Shield/Storm. The Army AH-1F was not equipped with FLIR, and the pilots flew with NVGs during the hours of darkness. The featureless desert and moonless nights proved too great a risk, however, and therefore the Army AH-1s were restricted to daylight operations. Generally assigned to air cavalry units, the Cobras performed well, leading the charge into Iraq and destroying several Iraqi armored vehicles and bunkers. In one instance an Army Cobra pilot, with his rockets expended and 20-mm cannon inoperable, made several dummy passes at Iraqi positions to draw fire and keep the enemy's heads down while his scout and wingman withdrew from an area where they had come under heavy ground fire.

The Marines deployed forty-eight Super Cobras to the Persian Gulf to support Operation Desert Shield/Storm. Despite the adverse environmental conditions of temperatures of 135 to 145 degrees Fahrenheit, blowing sand, and choking clouds of smoke from burning oil wells, Marine Cobras averaged a mission ready rate of 92 percent, better than that of any other attack helicopter in the Persian Gulf War. The Marine Cobras flew racetrack patterns behind friendly lines and made firing passes to support Marines on the ground, sometimes close enough that the empty 20-mm shell casings fell on the advancing Marines. On the third day of the war a Marine colonel, flying an FLIR-equipped UH-1N, won a Navy Cross by leading twelve AH-1Ws through the intense smoke of the burning oil fields to support the breakthrough to Kuwait City. During the "100 hour war" Marine Cobras destroyed at least 97 tanks, 104 armored personnel carriers and vehicles, 16 bunkers, and 2 antiaircraft artillery positions, without the loss of an aircraft.

With the AH-64 going into full production and in accordance

with a new Force Modernization Plan, the U.S. Army began to retire its fleet of AH-1 Cobras. In March 1999 the 25th Infantry Division (Light) in Hawaii, the last active-duty unit to employ the AH-1F, retired its Cobras. As of April 2001, 395 AH-1Fs remained in service with Army Reserve and National Guard units, but the Army continued with a program to place its remaining Cobras in long-term storage by fiscal year 2002.

Although the Army planned to phase out the AH-1s, the USMC moved to extend the service life of its Cobras. In the mid-1990s the USMC began to rectify several AH-1W shortcomings discovered during Desert Storm. To enhance the "Whisky" Cobra's night fighting capability, the corps contracted Bell Helicopter to retrofit its AH-1Ws with the Kollsman Night Targeting System (NTS), which comprised FLIR, low-light TV, a laser designator/rangefinder, and automatic target tracking. By 1996 the AH-1W NVG-compatible cockpits received enhanced avionics and electronics, including a tactical navigation system that incorporated a GPS with an imbedded inertial navigation system. The new EW countermeasures, in addition to the APR-39 Radar Warning Receiver, included a missile warning system that automatically signaled the AN/ALE-39 dispenser to fire chaff and flare decoys; the AN/AVR-2 laser detector warned the crew if a laser rangefinder was directed toward the helicopter. In 1998 the USMC took delivery of seven newly manufactured AH-1W Cobras, bringing its total inventory to 201.

Pilots of the Marine Air Wing participating in OIF used tactics similar to those of the first Persian Gulf War. USMC attack helicopter pilots never hovered but maintained a racetrack or similar pattern near the advancing Marine Ground Contact Elements (GCE) and dove in to provide Close Air Support (CAS) when requested. The Marine AH-1Ws fired all their weapons from running fire patterns, sometimes within only a few meters of the Marines on the ground. Marine Cobra pilots flew more than 2,500 combat hours in the first thirty days of the war, sometimes spending twelve to fourteen hours in the cockpit. Nearly all the Marine AH-1Ws received some type of combat damage, but mechanics had most of them back on operational status within twenty-four hours. Remembering the Army lessons of Mogadishu and Karbala, the Marines did not allow their Cobra pilots to hover in an urban environment, and as a result they avoided Iraqi "helicopter traps."

In August 1998 the USMC began a program to upgrade its Cobras to the next generation of attack helicopters by delivering four AH-1Ws to the Bell plant for conversion into AH-1Z test aircraft.

The previous month Bell had selected Lockheed-Martin to supply the "Hawkeye" Target Sight System (TSS) for the new Cobra. The TSS provided a third-generation thermal sighting system, an eye-safe laser rangefinder/designator, and a fully integrated computerized fire control system that allowed the crew to passively survey the battlefield, and track multiple targets, well beyond the maximum range of the aircraft's weapons. The newly designated AH-1Z Viper's cockpit contained the latest in digitized MFD instrumentation and advanced avionics, including a moving map display for both pilots. AH-1Z modifications included a new Model 680 bearingless composite four-bladed main rotor, a new transmission, a four-bladed tailrotor and drive train, a strengthened tailboom, and a more robust, energy-absorbing landing gear. In the same program, the Marines plan to modernize their fleet of UH-1Ns, which will be redesignated UH-1Y, with the same cockpit instrumentation and navigation systems, engines, and rotor systems. In addition to increased payload and maneuverability, the commonality of aircraft systems and almost identical cockpit stations will reduce maintenance costs and time to qualify pilots in both aircraft.

In December 2000 the first AH-1Z took flight, with the Marines expecting to receive 180 Vipers in a measured production run scheduled to end in 2013. Two General Electric T700-GE-401 1,690-horsepower engines powered the first test aircraft, but the more powerful T700-GE-701C 1,940-horsepower turboshafts may be installed in production models. In addition to the TSS, the Viper's updated weapons systems included the THALES Avionics "TopOwl" Helmet Mounted Sight and Display System (HMSD), a fire control system for the AGM-114A/B/C Hellfire antitank, AGM-114F Hellfire antiship, and AGM-114K Hellfire II radio-frequency antitank missiles, Mk 77 firebombs or cluster bomb units, and the Longbow Cobra Radar System (CRS). The CRS, contained in an EW wingstore pod, gave the Viper the same capability as the Apache Longbow to discriminate between friendly and enemy vehicles on the battlefield. The Viper retained the 20-mm cannon and missile racks for the Sidewinder and Sidearm missiles that became standard on the AH-1W. The refined armament system allowed the AH-1Z to accomplish the full spectrum of attack and reconnaissance missions, to include defeating a broad array of enemy threats at extended ranges. To extend the normal range of the aircraft, ground crews could also install 100-gallon auxiliary fuel tanks on the wings. Capable of nearly 200 knots, the Viper prototype had a maximum gross weight of 18,500 pounds, 10,000 pounds more than the original AH-1. The AH-1Z should reach a range of 700 miles carrying eight Hellfire missiles, fourteen Hydra 70 FFARs, and 650 rounds of 20-mm ammunition at 3,500 feet and 91 degrees Fahrenheit.

The U.S. Army deployed 300 AH-64s to Desert Shield/Storm for the "Mother of All Battles." Despite some shortcomings, the Apaches lived up to their crews' boast that they "ruled the night." As part of Desert Shield the first Apache unit arrived in Southwest Asia in November 1990, with additional units arriving until the commencement of Desert Storm. On January 18, 1991, Apaches from the 101st Airborne Division punched an 80-mile-wide hole in Iraqi air defenses that allowed USAF F-117s, F-111s, and other coalition fighter-bombers to appear undetected in the skies over Baghdad. Prior to the mission, ground crews replaced one rocket pod with an external wingtank on sixteen AH-64s, to give them the three-hour endurance required for the mission. Because of the lack of terrain references along the mission route, a USAF Sikorsky MH-53J Pave Low helicopter, using its ultrasophisticated navigation systems, served as pathfinder and led eight selected Apaches to their firing positions. At 0238 the Apaches' Hellfire missiles and 30-mm cannon fire obliterated a series of radar and communication installations, leaving the Iraqi air defenses blind to the incoming aerial onslaught.

Desert sand, however, played havoc with the AH-64 during the conflict. Sand the consistency of talcum powder sifted into everything, causing failures in fire control computers, air-conditioning systems, and—as with all helicopters in the theater—destroying rotor blades in just a few hours of flight time. Despite the heroic efforts of maintenance personnel, at least one-third of the AH-64s remained inoperative at any one time during the war. Although the Apache pilots could live in the desert with the infantry, their aircraft could not. The AH-64 averaged five hours of maintenance for each hour flown, the worst of any aircraft in the Gulf region. Apart from maintenance problems, however, the Apache pilots accomplished every mission, destroying hundreds of Iraqi tanks and armored vehicles. In several engagements Iraqi tankers abandoned their vehicles in the midst of a firefight because they could not discern what was killing one tank after another. Iraqis refused to sleep in their tanks at night, because they never knew when an undetected Apache might blast their vehicle with a Hellfire. In the last hours of the conflict Apache crews reported Saddam Hussein's vaunted Republican Guard, fleeing Kuwait on the "highway of death," indiscriminately firing their Soviet-made ZSU-23-4 and SAM antiaircraft systems into the air, vainly hoping to protect themselves from unseen Apaches. During Desert Storm the United States lost only one AH-64 shot down, and that probably because the pilot climbed too high, allowing enemy gunners to detect his aircraft. Although that aircraft was shot down, the crew survived and was rescued with no injuries. Iraqi troops became so fearful of U.S. firepower that they sometimes surrendered en masse to unarmed OH-58 scout helicopters.

Apaches continued to serve in Southwest Asia after Desert Storm. On April 24, 1991, eighteen AH-64s of the 6th Cavalry began a lengthy self-deployment from Germany to the Persian Gulf. The Apaches joined the task force providing security in northern Iraq as part of Operation Provide Comfort. By the late 1990s the U.S. Army had deployed AH-64 battalions in Germany and South Korea, as well as the continental United States.

In 1997, Boeing Rotorcraft acquired McDonnell-Douglas Aircraft and began marketing all McDonnell-Douglas products under the Boeing name. Although the name change was not a factor, the Apaches did not enjoy the same reputation and success flying in Bosnia and Albania as they had in Iraq eight years earlier. When the U.S. military finally decided to send AH-64s to support UN operations in the Balkans, as part of Task Force Hawk, twenty-four Apaches were sent to fight in the war with Yugoslavia. A series of nighttime accidents, and the over-reactions of uninformed reporters, unfortunately caused casualty-averse commanders to restrict the use of Apaches in the gloomy mountains after sunset.

After the AH-64A became operational, continued rigorous testing exposed several deficiencies in the Apache that were confirmed during combat operations in Panama and the Persian Gulf. Although more effective than conventional optics, FLIR resolution deteriorated during periods of "IR crossover," when the environment and a target's IR emissions were virtually equal. Common in all IR systems, crossover usually occurred near daylight or sundown, causing targets literally to disappear from the FLIR screen. Atmospheric conditions, especially moisture, compounded crossover problems. Low clouds also affected the Hellfire seeker head, preventing or breaking a target lock-on by defusing the reflected laser light. Dust, haze, rain, snow, smoke, and other battlefield obscurants diminished the effectiveness of the PNVS, TADS, and Hellfire seekers. Lengthy navigation over water degraded the reliability of the Doppler navigation system. Multiple rocket firings caused engine surges, restricting the pilots to firing rockets in pairs only. These shortcomings negated the AH-64's greatest strength, the ability to fly at 80 to 120 knots at treetop height and arrive at a decisive place and time in a battle without an enemy's knowledge.

To eliminate the shortcomings, the U.S. Army began testing the upgraded AH-64D Longbow Apache (LBA). The major improvement was the AN/APG-78 millimeter wave Fire Control Computer (FCC), which detected, classified, and prioritized fixed and moving targets, whether vehicles or low-flying aircraft. In addition, the FCC provided the pilot with obstacle detection for flying in adverse weather. The revamped Apache also included an imbedded GPS and inertial navigation system with an accuracy of less than 3 meters. With two target acquisition systems the AH-64D crew could search for targets independently, the pilot with the FCC and the copilot/gunner with the TADS.

The Army fielded the AH-64D in two versions, one with the FCC and upgraded T700-GE-701C engines and the second without the FCC and new engines, but with the capability to receive digitized target information from the FCC and fire the new generation of Hellfire missiles. A new AGM-114K Hellfire II Radio Frequency (RF) guided Hellfire missile, which was impervious to IR guidance problems, enhanced the Apache's ability to engage threats in adverse weather and through battlefield obscurants. All AGM-114 variants may be guided by either of two acquisition modes, lock-on-before-launch (LOBL) and lock-on-after-launch (LOAL), allowing engagements of ground and rotary-wing targets at more than 7 kilometers. In addition to the Hellfire, the AH-64D may be armed with AIM-9 Sidewinder AAMs, AGM-122 Sidearm antiradiation missiles, and the Advanced Precision Kill Weapon System (APKWS), an improved version of the Hydra 70 FFAR. In 1995 Operational Tests (OT), the AH-64D prototypes proved more than 400 percent more effective than the AH-64A, detecting, prioritizing, and engaging 128 armor, air defense, and countermeasure threats.

In October 1998 the 1-227 Aviation Battalion at Fort Hood, Texas, became the first operational AH-64D unit. The Army planned an extended remanufacture program to bring all existing AH-64A models to the Longbow standard, which should eventually number 936 AH-64Ds in twenty-five battalions assigned to active duty, reserve, and Army National Guard units. Each AH-64D company will be assigned three aircraft with the FCC and five without. All new and remanufactured AH-64Ds included APR-39 radar detectors, AN/AlQ-144 IR countermeasures set, AN/ALQ-136 radar jammers, and chaff/flare dispensers as an integral countermeasures system.

Future retrofits include Lockheed Martin's new Arrowhead FLIR system, which features improved resolution and target tracking. Israel, Egypt, Saudi Arabia, Kuwait, the United Arab Emirates (UAE), and Greece ordered export versions of the AH-64D.

In 2002 the United States deployed AH-64Ds to Afghanistan, where critics asserted that the Apaches would suffer the same ignominious defeat that befell Soviet helicopters a decade earlier. Although the high altitudes limited both the fuel and ammunition that the AH-64D could carry, the Apache and its crews refuted the naysayers. During Operation Anaconda, which took place in early March along the high ridgelines around the Shah-e-Kot valley, the AH-64s high gross weight and underpowered engines precluded the aircraft from hovering, forcing Army pilots to relearn the virtues of running fire. When al-Qaida forces pinned down a battalion of U.S. infantry, the Apaches flew through witheringly heavy machine gun fire and volleys of RPGs and, "like fire-spitting avenging angels," blasted the dug in al-Qaida with rocket and 30-mm cannon fire (Naylor 2002, 15).

The seven Apaches involved in the fight on March 2 saved the beleaguered infantry but sustained extensive combat damage that knocked four of the helicopters out of the fight. Hit by an RPG, one Apache lost all of its engine oil. Chief Warrant Officer 4 James "Jim" Hardy radioed Chief Warrant Officer 3 Keith Hurley, flying the stricken aircraft to land; among whizzing enemy bullets, the two pilots poured six quarts of spare oil into the hemorrhaging Apache. Hardy then jumped into the damaged AH-64 and flew it more than 50 miles to safety, out of reach of al-Qaida. When maintenance crews examined the AH-64s, twenty-seven of the twenty-eight main rotor blades had bullet holes in them, and one aircraft sported an additional thirteen bullet and shrapnel holes (Naylor 2002, 15).

When U.S. Army Apaches reached Iraq in 2003, during Operation Iraqi Freedom (OIF), or the Second Gulf War, in an attempt to recover a reputation besmirched in Bosnia, overzealous commanders committed serious errors in employing the AH-64. On the night of March 23, 2003, thirteen AH-64Ds from the 6th Cavalry Regiment and another eighteen Apache Longbows from the 227th Aviation Regiment took off on a deep strike mission to attack armored columns of the Iraqi Republican Guard's Medina Division near Karbala. Overconfident of the Longbow's capabilities, and ignoring lessons learned in Somalia concerning volleys of RPGs mixed with small arms fire, U.S. commanders allowed themselves to be lured into a "helicopter trap" by Iraqi intelligence. Iraqi agents watched the U.S. helicopters lift off and alerted their compatriots by cell phone. One overloaded Apache crashed on takeoff, but the crew was uninjured. The remaining AH-64Ds flew into a firestorm of everything from S-60 57-mm AA guns to individuals with AK-47s. With tracers crisscrossing the sky and flak bursts as big as their helicopters, one unit turned back before reaching its attack positions; another managed to fire only a few missiles before the intense ground fire drove it off as well. On the return trip small arms fire downed one AH-64D, but again, attesting to the Apache's ruggedness, the crew survived unhurt. The two pilots were, however, captured when their wingman exhausted his 30-mm ammunition and had to flee before being shot down. Although all the Apaches received hits, only one crewmember required hospitalization, and all but five of the damaged AH-64s were back in service within twenty-four hours.

Since the days of Vietnam, attack helicopter instructor pilots have taught their fellow aviators that, if terrain does not provide masking (cover) for the helicopter, only high-energy (airspeed) firing passes provide any hope of survival. In the initial engagements in Iraq, several U.S. commanders wanted the Apache crews to conduct hovering fire from suburban areas. The housetops provided perfect firing positions for the fedayeen and Iraqi irregulars, who laced the vulnerable helicopters with a withering fusillade of small arms fire. Despite all the high-tech antiaircraft weapons, an individual with an automatic rifle or RPG remained the greatest threat to hovering helicopters.

On successive missions, Army commanders adopted the tactics used by the 101st Airborne Division. The "Screaming Eagles" integrated their aviation assets with advancing ground forces into a combined arms team, supported by conventional tube artillery and Army Tactical Missile Systems (ATACMS), a Forward Air Controller (FAC) with direct communications with USAF CAS aircraft, JS-TARS aircraft to provide indications of any enemy vehicles moving in the target area, EA-6B or other EW aircraft to jam enemy radar and communications, and Unmanned Aerial Vehicle (UAV) support to provide real-time video of the target area. The 101st Airborne Division also sent small reconnaissance units to move forward to find and fix enemy positions, then the Apache battalions circled to the flank and rear of the enemy to attack from an unexpected direction. Following lessons relearned in Afghanistan with the overweight and underpowered AH-64Ds, 101st attack helicopter pilots maintained their airspeed, using running fire techniques to employ their Hellfire missiles as well as rockets and cannon. Even though they were flying at much lower altitudes than in Afghanistan, temperatures in excess of 110 degrees Fahrenheit made hovering out of the question. Making use of these tactics, Apache units in CCA destroyed more than 800 targets in five days without losing a single aircraft. On one mission alone, Apaches destroyed forty Iraqi tanks, APCs, and artillery pieces, while sustaining a single enemy round through a rotor blade of one AH-64. After adopting these tactics the United States lost no other AH-64s in Iraq to enemy action, although several were damaged by ground fire and at least three crashed taking off or landing in the horrific "brown-outs" of blowing dust whipped up by their own rotorwash.

On September 25, 1998, Westland responded to the British Army's requirement for a new attack helicopter by producing the prototype WAH-64 Apache under license from Boeing. On July 31, 2000, Westland GKN delivered its first AH-1 Mk 1 to the Ministry of Defense, and on January 16, 2001, the director of British Army Aviation signed the release for service for the first nine of sixty-seven AH-1 Mk 1 Apaches ordered by the British Army. Eventually the Apache will replace the TOW-equipped Lynx Mk 7 in the British Army.

Two Rolls-Royce Turbomeca RTM322 2,100-horsepower turboshafts powered the 47-foot, 10-inch four-bladed composite main and tailrotors. At a designed speed of 150 knots the Mk 1 Apache attained a range of 750 nautical miles with available ferry tanks. Equipment included the latest in technological advances, including the Longbow fire control radar, Hellfire missiles, both semiactive laser and radio frequency versions, CRV7 ground suppression rocket system, and the standard AH-64 30-mm cannon.

In the final years of the twentieth century attack helicopters, as with combat support rotorcraft, received unstinted time and funds, but generally from consortiums and not from individual manufacturers. A Franco-German syndicate jointly produced an attack helicopter called the Tiger. Aerospatiale and MBB actually began developing the machine in 1988, and when the companies merged Eurocopter France and Eurocopter Deutschland each owned 50 percent of Eurocopter Tiger GmbH, the primary contractor for the Tiger, originally the PAH-2. Envisioned as an all-weather two-pilot *PanzerAbwehr Hubschrauber,* combat helicopter, to fulfill antitank and escort/support roles, the PAH-2 differed from most attack helicopters in that the pilot sat in front and the copilot/gunner behind in a tandem seating arrangement. The company constructed the helicopter around advanced avionics, an "all glass cockpit," multiple weapons systems, including the Euromissile HOT and TRIGAT "fire and forget" ATGM, and stealth technology. The first Tiger took to the air in April 1991, and the second, in *Gerfaut* (escort) configuration, on April 22, 1993. In 2001 both the French ALAT and German Army accepted the first of 427 Tigers ordered by the two countries.

The Tiger featured stub wings with dual hardpoints for two 12-shot, 68-mm FFAR, rocket pods, up to eight ATGMs, and two Mistral AAM, and a chin turret mounting a 30-mm cannon. The weapons system included either a roof- or mast-mounted sight with infrared, TV, and conventional optics, along with advanced fire control computers for maximum accuracy. An air-to-air package of 12.7-mm gun pods and Stinger missiles was also available. An FLIR system provided the pilot with imagery for flying in low light and restricted visibility. Countermeasures included radar/laser/missile warning receivers and flare and chaff dispensers.

Two MTU/Turbomeca/Rolls-Royce MTR.390 1,280-horsepower turboshafts, equipped with high-tech IR suppressors, powered a four-bladed main rotor and a three-blade tailrotor mounted on a large vertical fin. Engineers made use of composite materials in the fuselage and rotor systems, and extensive nonmetallic armor plating, to lighten the helicopter and improve its battlefield survivability, including operations in a chemical, biological, and radiological (CBR) environment. The Tiger exhibited a range of 432 nautical miles at a cruise airspeed of 150 knots. In December 2001 the Australian Army ordered twenty ARH Tigers, a scout/reconnaissance version of the helicopter. Australian Aerospace, a division of Eurocopter, received a contract to assemble these aircraft locally, modified with a laser designator to fire Hellfire antitank missiles. The Australian Army accepted the first deliveries in 2004. In 2003 Spain also ordered twenty-five Tigers, with upgraded engines and deliveries scheduled in 2007 and 2008.

Along the Pacific Rim another armed helicopter took shape. On August 6, 1996, the first prototype of the XOH-1 Kongata Kansoku light observation helicopter made its initial flight in Japan. Produced by a Japanese consortium headed by Kawasaki Heavy Industries and designed to replace the OH-6 in service with the JGSDF, the helicopter's name belied its appearance, which resembled the tandem-seated Eurocopter Tiger and Denel Rooivalk. With a 38.058-foot, four-bladed hingeless, composite main rotor and eight-bladed Fenestron, along with armament hardpoints on its stub wings, the OH-1 looked more like an attack helicopter than an armed reconnaissance aircraft. Known as the Ninja since August 2000, the OH-1 had two Mitsubishi XTS1-10 885-horsepower turboshafts, with a thirty-minute "dry run" capability. Instrumentation included a computer integrated cockpit and flight direction system that afforded the helicopter with both day and night capabilities. Equipped with FLIR and a laser designator/rangefinder linked to a fire control computer, the Ninja had typical armament of FFAR rocket pods, TOW ATGM, or AAMs. With a maximum takeoff weight of 8,820 pounds, it attained an estimated airspeed of 145 knots and a range approximating 300 nautical miles. The consortium delivered the first OH-1 armed reconnaissance version in 2000, and hoped to sell the Ninja to the JGSDF as an attack helicopter as well; the Japanese military, however, selected the U.S. AH-64D.

In the early 1990s the Russian military decided that it needed a two-seat version of the Ka-50 Hokum. The flying prototypes of the Ka-50A and 50N indicated that a two-pilot aircraft was needed as a day/night, all-weather close support attack helicopter, and that such an advanced machine demanded a training variant. As a result, in 1997, Kamov introduced the Ka-52 Hokum B, dubbed the Alligator by the Russians. The Ka-52's dimensions and flight characteristics remained virtually the same as those of the Ka-50, but with a widened cockpit to accommodate two pilots sitting side by side. The additional weight degraded performance, but to no significant degree. The Ka-52 possessed the same TV, thermal, laser, and optical sighting systems linked to identical fire control and navigation computers installed on the Ka-50N, but it also included a state-of-technology mast-mounted millimeter wave radar to identify multiple enemy threats. Armament consisted of the fixed 30-mm cannon and a mix of AT-9 Vikhr ASMs, S-8 80-mm FFARs, and AA-11 Archer AAMs. Kamov sought to sell the Alligator outside of Russia, with the French companies Thomson-CSF and Sextant Avionique offering navigation and attack electronic systems for the export variants of the Ka-52.

In November 1999, Kamov partnered with Israeli Aircraft Industries (IAI) Corporation to bid for a 145 aircraft contract for the Turkish Army. The combine offered the Ka-50-2, a tandem seating version of the Hokum, to meet the requirements for a helicopter gunship demanded by the contract. In 2000 the Turkish government eliminated Boeing and Eurocopter from the competition, leaving the Kamov/IAI Ka-50-2 Erdogan, the Agusta A129 Mongoose, and the Bell-Textron AH-1Z King Cobra in contention for the lucrative contract.

The world's most advanced reconnaissance/attack helicopter can trace its antecedents to the Light Helicopter Experimental (LHX) combat helicopter program, initiated in 1983 by the U.S. Army to replace the UH-1, OH-58, and AH-1 with a single helicopter. Envisioned at the time to involve as many as 6,000 helicopters, discoveries during R&D and funding reduced the number to about 1,200 aircraft, and then only a reconnaissance/attack version. Hoping to create the most advanced helicopter on the battlefield, the Army intended to maximize the use of stealth technology and composite construction in the LHX, and to provide its crew with new sensors, millimeter wave radar, computerized weaponry, and advanced night vision systems.

In 1984, Bell Helicopter proposed its lightweight Bell Advanced Tilt Rotor (BAT) for the LHX. The single-seat, tilt-rotor aircraft had to be discounted, however, on account of weight. LHX requirements restricted maximum gross weight to 7,000 pounds, and the BAT exceeded that mark. In June 1985, Sikorsky and Boeing teamed to compete for the LHX contract. To meet LHX requirements, originally conceived as a single-pilot aircraft, Sikorsky modified a S-76 to study the human engineering required for the LHX. Sikorsky's investigations concluded that, even with advanced computer technology, a single pilot configuration would overload the pilot in a combat situation. As a result of Sikorsky's conclusions, the Army revised the LHX to a two-place helicopter.

In June 1988 the Department of Defense issued an RFP, and two "teams" presented designs for evaluation—the Boeing-Sikorsky LH First Team and the Bell-McDonnell Douglas LH Super Team. In January 1991 the Army selected the Boeing-Sikorsky LH First Team to produce the first two prototype LHX helicopters, redesignated the RAH-66 Comanche. Boeing and Sikorsky used almost 100 percent composites in constructing the Comanche, a tandem two-seat helicopter. Two LHTEC T800 turboshafts drove the advanced 40-foot, five-bladed bearingless composite main rotor and eight-bladed shrouded tail fan. The aerodynamic 46-foot, 10-inch fuselage housed the retractable tricycle landing gear and a chin-mounted 20-mm cannon. The first RAH-66 prototype rolled out on May 25, 1995, and made its initial flight on January 4, 1996. By 1998, however, the Comanche had completed only 105 hours of test flights because of a paucity of funding.

The Comanche's composite stealth technology, advanced electronic sensors, and computerized helmet and cockpit displays made the helicopter one of the deadliest in the world. The strength of the RAH-66's composite construction allowed the pilot, flying at 100 knots, to turn the nose any direction and engage targets while continuing on the same flight path. The advanced rotor system reduced the Comanche's acoustic signature, and hidden exhausts defeated all heat-seeking missiles. To further reduce its radar cross-section, the aircraft carried Hellfire, or more advanced, missiles in internal weapons bays. The suite of night vision sensors/devices and GPS navigation system gave the Comanche a true day/night all-weather capability. To reduce maintenance time and costs the manufacturer installed more than 1,800 integrated self-test diagnostics. Despite advanced technology and successful tests, however, DoD canceled production of the Comanche in 2004, instead buying several more versions of the Blackhawk, Chinook, Kiowa Warrior, and Apache already in production.

TILT ROTOR AIRCRAFT

In the mid-1980s the U.S. DoD funded comprehensive evaluations of the Bell-Boeing XV-22 Osprey, which melded the vertical lift of a helicopter with the cruise speed of a fixed-wing turboprop aircraft. The program called for extensive use of composite materials in order to construct a lighter and stronger aircraft. A progeny of both the Boeing VZ-2 and the Bell XV-15, the V-22 featured two Allison T406 6,140-horsepower turboshafts installed in the tips of slightly forward-swept wings, powering twin counter-rotating, three-bladed composite 36-foot proprotors. The swiveling engine nacelles allowed the Osprey to lift off to a hover and accelerate into forward flight in less than twelve seconds. A sturdy retractable landing gear supported the aircraft for ground operations.

Initially the U.S. Army managed the JVX project, but in January 1983 the U.S. Navy assumed stewardship of the program. In June 1985, DoD let a contract for six flying prototypes and several static test mock-ups. The U.S. military services anticipated a requirement for 913 V-22s of several variants. To replace its CH-46 fleet, the USMC expected to order 552 MV-22A assault aircraft, each capable of transporting up to twenty-four combat-loaded Marines. The Army wanted 231 similar aircraft. The USAF foresaw a requirement for eighty CV-22 special operations long-range transports. Along the same lines the Navy submitted a need for fifty HV-22A aircraft for SAR, special warfare, and fleet logistical support. The Navy also expected to order an additional 300 Ospreys equipped for ASW. For shipboard storage, engineers designed the Osprey's wings to fold fore and aft along the fuselage, with the folded blades and engine nacelles rotating and folding parallel to the wings and fuselage.

On March 19, 1989, after a long series of simulated test-stand flights, the first V-22 lifted off vertically for a twelve-minute flight. The Osprey completed its first transition from a hover to horizontal flight on September 14. By May 9, 1990, two additional machines had completed their test flights. In June 1991 the fifth XV-22 sustained considerable damage in a nonfatal crash on its first flight. A more calamitous event occurred on July 21, 1992, when the fourth ship crashed, killing all seven aboard. The Bell-Boeing team experienced a number of vibration and engine fuel control problems that caused the XV-22 program to fall well behind schedule. Additionally, a funding fight between DoD and Congress created a shortage of funds that precluded completion of the sixth aircraft; its parts were used as spares for the other five prototypes. Although only the USMC remained in the program, Congress finally agreed to fully fund the MV-22A Osprey in September 1994. Five aircraft, representing production models, began test flights on February 5, 1997. The USMC received its first Osprey in 1999 and began operational testing. A series of accidents, however, including one in April 2000 that killed nineteen Marines, again halted the program for a short period. A USMC accident report indicated that pilot error had caused the April 2000 crash, when the pilot inadvertently entered a condition called vortex ring state. In that situation a rotor system continues to recirculate disturbed air, causing the rotorcraft to settle in its own rotorwash.

The MV-22A's 57-foot fuselage carried up to twenty-four troops, or twelve litters plus medical attendants, or 20,000 pounds of internal cargo, or a 15,000-pound slingload. The fuselage had a large rear loading ramp to facilitate handling or large and wheeled cargoes. The Osprey cruised at 100 knots in the helicopter mode and 315 knots in the airplane configuration. The aircraft had a service ceiling of 26,000 feet and a range of 500 nautical miles without auxiliary fuel. The V-22 could also be refueled by its air-to-air refueling probe. If the MV-22A projects continue as planned, the USMC will receive 425 of the aircraft.

NOTAR

One of the most radical innovations in helicopter design began in the mid-1940s. In 1944, James Weir of the Cierva Autogyro company in Britain produced the W-9 helicopter. The W-9 appeared much like other early helicopters, with a three-bladed main rotor and bubble canopy, but the tubular tailboom had no tailrotor. Instead exhaust gases ducted into the tailboom and ejected through nozzles replaced the antitorque rotor. The W-9 first flew in 1944, but it crashed in 1946 and was abandoned—but not before laying the foundation for the modern NOTAR ("No TailRotor" concept).

In 1962, Hughes Helicopters rejoined the U.S. Army investigations into rotors powered by pressure-jets. Hughes constructed the Model 385, designated XV-9 experimental helicopter by the Army. On November 5, 1964, the XV-9 made its first flight. The helicopter, fabricated from the nose of a Hughes 369, an H-34 landing gear, and components from other aircraft, was powered by two General Electric YT64 gas generators. Installed externally on pods, the generators produced hot exhaust gases that were ducted to pressure nozzles at the tips of the three main rotor blades, providing thrust to lift the machine.

In the late 1970s and early 1980s, Hughes pursued the idea of controlling a helicopter with pressure-jets but changed the focus to the antitorque system, initiating the NOTAR project. McDonnell-Douglas, and then Boeing, continued the program after acquisition of the Hughes company. Several aerodynamic engineers believed the NOTAR to be a more efficient and less vulnerable antitorque system than the traditional tailrotor. They replaced the existing Model 500 tailboom with a larger tubular tail section, from which pressurized air was ejected through vents installed at the end of the boom to counteract the torque created by the main rotor system. Although the military expressed little interest, Hughes continued to research the NOTAR thrust system and, on December 17, 1981, an OH-6 modified with the NOTAR boom took off on its maiden flight.

After McDonnell-Douglas obtained Hughes Aircraft, the company decided to offer the system on McDonnell-Douglas civil helicopters. In December 1989 the prototype MD 530N, essentially, a Model 530K with the NOTAR tailboom, made its first flight. After certification by the FAA, McDonnell-Douglas marketed several larger versions of the NOTAR design, the first being the MD 520 N, with a larger cabin and a five-bladed main rotor powered by a single Allison 250-C47M 800-horsepower engine. Successive NOTAR models introduced into the late 1990s included the 600N and 900N series of executive helicopters with luxurious club seating for six to eight passengers. The latest versions featured larger five- or six-bladed main rotors and options of twin engine packages such as two 629-horsepower Pratt & Whitney PW206B turboshafts. With the

NOTAR designs Boeing claimed to produce some of the quietest and fastest executive helicopters on the market.

CONCLUSION

Since the introduction of helicopters into the Korean War in the 1950s, no conflict in the world has occurred without the employment of vertical rising aircraft. Becoming an integral, and sometimes decisive, component of modern military units, rotorcraft changed forever the art and science of warfare. Airmobile tactics employed in Vietnam, and refined by several countries over the years, allow commanders to influence the course of a battle rapidly by adding a flexibility of response not available before the advent of helicopters. To exploit airmobile concepts fully, manufacturers made use of the most powerful engines and advanced technology to create the most versatile combat assault and most lethal attack helicopters possible, all capable of operating day or night, in all weather conditions, in a high-threat environment. Problems of vibration, noise, and safety continued to plague many civilian operators because the traveling public demanded more comfort and luxury than soldiers, and qualitative changes in military technology solved many of the problems that beset civilian operators. Improvements in composite construction lightened the aircraft, thus increasing payloads, and stealth technology led to quieter helicopters.

Experimentation with innovations such as the Sikorsky advancing blade concept, Westland's BERP rotor blades, Hughes/McDonnell-Douglas/Boeing NOTAR helicopters, Bell's nodular attachments of fuselage to airframe, and wide use of composites produced strong, reliable helicopters well regarded throughout the world. The MV-22, for example, is constructed more than 70 percent of composite materials, as are most of the world's newly designed helicopters. Many international aviation experts believe that two designs are destined to dominate future rotorcraft: the classic helicopter and a "tilt something." Another innovation for the future battlefield is the Horizontal or Vertical Take Off or Landing (HOVTOL) UAV. A May 22, 2000, *Time* magazine article stated: "By 2025, the U.S. will use fleets of UNMANNED combat air vehicles to attack missile batteries before they can target troop-carrying planes and helicopters." Many of these new reconnaissance and attack UAVs may well be the helicopter or tilt-rotor designs now being tested. These same UAVs may support law enforcement, firefighting, SAR, weather data collection, or news services (*Time*, May 22, 2000, 98–100).

During the late twentieth and early twenty-first centuries, mergers in the aerospace industry created an almost homogeneous affiliation of international helicopter manufacturing corporations. With the collapse of the Soviet Union, military leaders and politicians alike had to alter their attitudes and perceptions of battlefields, and what weapons were required in the post–Cold War world. No longer did massed armor confrontations seem likely in Europe, and after 1991 the same appeared to be true for the Persian Gulf region. True, the naval missions remained much the same, but land warfare required lighter, more mobile forces. Rotorcraft of all types met the requirements for rapid deployment, logistical support, close combat support, and survivability demanded on twenty-first-century battlefields. In the civilian world rotorcraft became even more essential for transportation, EMS, SAR, mineral exploration, agriculture, fire fighting, and any other endeavor that an ingenious pilot, operator, or manufacturer might contrive.

REFERENCES

Aircraft of the World: The Complete Guide. N.p./USA: International Masters Publishers AB, licensed to IMP, 1996.

Bell Helicopter-Textron, http://www.bellhelicopter.textron.com. Accessed June 2002.

Boeing Company, http://www.boeing.com. Accessed January 2002.

Bowden, Mark. *Black Hawk Down: A Story of Modern War.* New York: Atlantic Monthly Press, 1999.

British Army Air Corps Association, http://www.aacn.org.uk. Accessed May 2002.

British Army Air Corps Historical Flight, http://www.rdg.ac.uk. Accessed May 2002.

British Army Air Corps Museum, http://www.flying-museum.org.uk. Accessed May 2002.

British Army Official Site, http://www.army.mod.uk. Accessed May 2002.

British Helicopter Museum, http://www.hmfriends.org.uk. Accessed May 2002.

Brown, David A. *The Bell Helicopter Textron Story: Changing the Way the World Flies.* Arlington, TX: Aerofax, 1995.

Chant, Christopher. *Fighting Helicopters of the 20th Century.* Christchurch, Dorset, England: Graham Beehag Books, 1996.

Donald, David, ed. *The Complete Encyclopedia of World Aircraft*. New York: Barnes and Noble Books, 1997.

Endres, Gunter, and Michael J. Gething, comps. and eds. *Jane's Aircraft Recognition Guide*. 3d ed. New York: HarperCollins Publishers, 2002.

Fort Rucker, AL. Home of U.S. Army Aviation, http://www.-rucker.army.mil. Accessed April 2002.

Fredriksen, John C. *Warbirds: An Illustrated Guide to U.S. Military Aircraft, 1915–2000*. Santa Barbara, CA: ABC-CLIO, 1999.

Gordon, Yefim, and Dimitriy Komissarov. "Mil Mi–24 'Hind.'" *World Airpower Journal* 37 (summer 1999), pp. 42–84.

Gunston, Bill, and Mike Spick. *Modern Fighting Helicopters*. London: Crescent Books, 1996.

Helicopter World, http://www.helicopter.virtualave.net. Accessed April 2002.

Helicopter's History Site, http://www.helis.com. Accessed June 2002.

"Helicopters of the U.S. Army," http://www.geocities.com/capecanaveral/hangar/3393/army.html. Accessed August 2002.

Higham, Robin, John T. Greenwood, and Von Hardesty. *Russian Aviation and Air Power in the Twentieth Century*. London: Frank Cass, 1998.

Hirschberg, Michael, and David K. Daley. *U.S. and Russian Helicopter Development in the 20th Century*. N.p.: American Helicopter Society International, 2000.

Igor Sikorsky Historical Archives, Inc., http://www.iconn.net/igor/index lnk.html. Accessed October 2002.

International Helicopters, http://www.globalsecurity.org. Accessed August 2002.

Kelly, Orr. *From a Dark Sky: The Story of the U.S. Air Force Special Operations*. Novato, CA: Presidio Press, 1996.

Keogan, Joseph. *The Igor I. Sikorsky Aircraft Legacy: The Chronology of Fixed-Winged and Rotary-Wing Aircraft of Igor I. Sikorsky and the Sikorsky Aircraft Company*. Stratford, CT: Igor I. Sikorsky Historical Archives, 2003.

Kuznetsov, G. I. *Kamov OKB 50 Years, 1948–1998*. Edinburgh, Scotland: Polygon Publishing House, Birlinn Publishing, 1999.

McGuire, Francis G. *Helicopters 1948–1998: A Contemporary History*. Alexandria, VA: Helicopter Association International, 1998.

Naylor, Sean D. "In Shah-E-Kot, Apaches Save the Day—And Their Reputation." *Army Times,* March 25, 2002, p. 15.

O'Grady, Scott. *Return with Honor*. New York: Harper, 1996.

Pember, Harry. *Seventy-five Years of Aviation Firsts*. Stratford, CT: Sikorsky Historical Archives, 1998.

Ripley, Tim. *Jane's Pocket Guide: Modern Military Helicopters*. England: Jane's, 1997.

Royal Air Force Official Site, http://www.raf.mod.uk. Accessed May 2002.

Royal Navy Official Site, http://www.royal-navy.mod.uk. Accessed May 2002.

Russian Aviation Museum, http://www.ctrl-c.liu.se/misc/ram. Accessed November 2002.

Saunders, George H. *Dynamics of Helicopter Flight*. New York: John Wiley and Sons, 1975.

Sikorsky Aircraft Corporation, http://www.sikorsky.com. Accessed June 2002.

Simpson, R. W. *Airlife's Helicopters and Rotorcraft: A Directory of World Manufacturers and Their Aircraft*. Shrewsbury, England: Airlife Publishing Ltd., 1998.

Soviet/Russian Helicopters, http://www.royfc.com/links/acft_coll. Accessed November 2002.

Unusual Aircraft, http://www.unrealaircraft.com. Accessed March 2003.

U.S. Air Force Historical Research Agency, http://www.au.af.mil/au/afhra/. Accessed December 2002.

U.S. Air Force Museum, http://www.asc.wpafb.af.mil/museum. Accessed December 2003.

U.S. Army Aviation and Missile Command (AMCOM), http://www.redstone.army.mil/history/aviation/. Accessed February 2002.

Uttley, Matthew R. *Westland and the British Helicopter Industry, 1945–1960: Licensed Production vs. Indigenous Innovation*. London: Taylor and Francis, 2001.

Helicopter
Specifications

AEROSPATIALE ALOUETTE II, SE 3130, SA313
Courtesy of Art-Tech/Aerospace/M.A.R.S/TRH/Navy Historical

COUNTRY OF ORIGIN: France

CREW: 1 pilot and 2 passengers

ROTOR DIAMETER: 36 ft. 2 in.

LENGTH: 42 ft. 5 in.

ARMAMENT: Some models with machine guns, or SS-10/SS-11 wire-guided antitank missiles. French Navy versions with homing torpedo.

POWERPLANT: One 530-horsepower Turboméca Artouste, or Astazou IIA turboshaft

AIRSPEED: Maximum 110 knots

RANGE: 278 nautical miles

CARGO CAPACITY: 2 external litters, or 2,500-pound slingload

NOTES: All helicopters in this family are visibly similar and differ principally in powerplants. More than 1,500 produced. SA 315B "Llama" initially built for India, sold to operators in more than twenty countries. "Cheetah" is built under license by HAL for Indian Army. Also produced under license in Brazil and Romania.

AEROSPATIALE ALOUETTE III, SA 316-319
Courtesy of Art-Tech/Aerospace/M.A.R.S/TRH/Navy Historical

COUNTRY OF ORIGIN: France

CREW: 1 pilot

ROTOR DIAMETER: 36 ft. 2 in.

LENGTH: 42 ft. 1 in.

ARMAMENT: Some versions with 7.62-mm machine guns and at least one variant with a single 20-mm cannon. Provisions for carrying four AS-11 or two AS-12 antitank missiles.

POWERPLANT: 319B 870-horsepower Turbomeca; 319 Astazou XIV turboshaft

AIRSPEED: SA 316B cruise 115 knots and SA 319B cruise 120 mph

RANGE: 290 nautical miles

CARGO CAPACITY: Up to 6 passengers, or 1,650-pound slingload

NOTES: First flew in 1959 as SE 3160; later production aircraft designated SA 316B. Larger cabin, more powerful engine, and greater performance than Alouette II. Some 1,455 Alouette IIIs, of all versions, delivered to more than seventy countries when production ceased in 1985. Built under license in India, Switzerland, and still in production in Romania.

AEROSPATIALE SA 321 SUPER FRELON
Courtesy of Art-Tech/Aerospace/M.A.R.S/TRH/Navy Historical

COUNTRY OF ORIGIN: France
CREW: 2 pilots and crewchief
ROTOR DIAMETER: 62 ft.
LENGTH: 75 ft. 7 in.
ARMAMENT: 321 G/H, 4 homing torpedoes or 2 AM-39 Exocet missiles
POWERPLANT: 3 Turbomeca Turmo IIIC turboshafts
AIRSPEED: 148 knots
RANGE: 550 nautical miles
CARGO CAPACITY: 27 passengers, 11,023 pounds cargo

NOTES: First flight 1962, with three main variants, military and civilian transport, and ASW. Civilian models carry from 34 to 37 passengers. The 321G is flown by the French Navy and is equipped with search radar, torpedoes, and other ASW equipment. The 321H Army and Air Force variant carries from 27 to 30 troops or cargo. The Israeli Defense Force operates several Super Frelons.

AEROSPATIALE SA 341/342 GAZELLE
Courtesy of Art-Tech/Aerospace/M.A.R.S/TRH/Navy Historical

COUNTRY OF ORIGIN: France

CREW: 1 or 2 pilots

ROTOR DIAMETER: 34 ft. 5 in.

LENGTH: 39 ft. 4 in.

ARMAMENT: SA 342 L/M, four Mistral AAMs, AS-11/12, or HOT ATGM. 2.75-inch rocket pods.

POWERPLANT: 1 Astazou XIVM turboshaft

AIRSPEED: Maximum, 167 knots

RANGE: 360 nautical miles

Cargo: Up to 4 passengers, or 2 litters

NOTES: First flight on April 7, 1967; placed in production 1971. Eleven variants placed in service: SA 341 B-E for British armed forces; 341 F for French Army; SA 341 G and 342 J for civil operators; SA 341 H military export version; and SA 342 K-M military versions. Under a 1967 Anglo-French agreement, Westland Helicopters produced "Gazelles," and also under license in Yugoslavia, India, and Egypt. More than 1,400 produced.

AEROSPATIALE/EUROCOPTER AS 550/555 (AS 355 CIVIL VARIANT) ECUREUIL/FENNEC

Courtesy of Art-Tech/Aerospace/M.A.R.S/TRH/Navy Historical

COUNTRY OF ORIGIN: France

CREW: 1 or 2 pilots

ROTOR DIAMETER: 35 ft. 1 in.

LENGTH: 42 ft. 8 in. (rotors turning)

ARMAMENT: 2-TOW or HOT ATGMs, torpedo, FFARs, or 7.62/12.7-mm machine gun pods; one model with 20-mm cannon

POWERPLANT: 1 or 2 Turbomeca Arriel I/A turboshafts

AIRSPEED: 155 knots

RANGE: 362 nautical miles

CARGO CAPACITY: 2 to 4 passengers or 2,000-pound slingload

NOTES: Prototype Ecureuil (Squirrel) flown in June 1974; twin-engine model 355 in September 1979. Military versions produced after 1990 redesignated AS 550/555 Fennec. Also manufactured under license in Brazil, while China's CHAIC Z-11 is an apparent copy. More than 2,250 of all models produced and used worldwide.

AEROSPATIALE/WESTLAND SA 330 PUMA
Courtesy of Art-Tech/Aerospace/M.A.R.S/TRH/Navy Historical

COUNTRY OF ORIGIN: United Kingdom, France

CREW: 2 pilots and crewchief

ROTOR DIAMETER: 49 ft. 2 in.

LENGTH: 59 ft. 6 in.

ARMAMENT: Giat 20-mm cannon under nose; AAM, ATGW, 57- or 70-mm rockets

POWERPLANT: 2 Turbomeca Turmo IvC turboshafts

AIRSPEED: 158 knots

RANGE: 309 nautical miles

CARGO CAPACITY: 20 passengers, or 7,055-pound slingload

NOTES: Originally a French product that first flew in 1965, the Puma became a joint Anglo-French helicopter. The RAF and French Army make use of the aircraft as a transport, and more than 700 were sold to various countries for both military and civilian users. The Puma carries 16 troops or up to 20 passengers in some civilian configurations. Current models are the 330J for civilian use and the 330L for military. Built under license in Romania.

AEROSPATIALE/EUROCOPTER 332 SUPER PUMA/COUGAR
Courtesy of Art-Tech/Aerospace/M.A.R.S/TRH/Navy Historical

COUNTRY OF ORIGIN: France

CREW: 2 pilots and crewchief

ROTOR DIAMETER: Cougar Mk II 53 ft. 1 in.

LENGTH: 63 ft. 7 in.

ARMAMENT: Combinations of 20-mm cannon, Exocet ASM, homing torpedoes, or 7.62/12.7-mm machine gun pods

POWERPLANT: 2 Turbomeca Makila 1A2 turboshafts

AIRSPEED: 170 knots

RANGE: 650 nautical miles

CARGO CAPACITY: 31 troops, or 9,920 pounds cargo

NOTES: The much modified Super Puma (redesignated Cougar in 1990) has a lengthened nose, more powerful engines, and a sturdier landing gear. Built in five versions, the 332B military version seats 20 troops; the 332F is equipped for ASW, SAR, and other naval missions; the "L" and "M" models have an extended fuselage allowing for four more seats. Bristol Helicopters operate 35 332L variants.

AEROSPATIALE/EUROCOPTER SA 360 AND SA 361 H DAUPHINE
Courtesy of Art-Tech/Aerospace/M.A.R.S/TRH/Navy Historical

COUNTRY OF ORIGIN: France

CREW: 1 or 2 pilots

ROTOR DIAMETER: 37 ft. 8 in.

LENGTH: 43 ft. 3 in.

ARMAMENT: 7.62-mm machine guns

POWERPLANT: SA 360 one 1,050-horse-power turboshaft; SA 361 H 1,400-horsepower turboshaft

AIRSPEED: Cruise, 150 knots

RANGE: 430 nautical miles

CARGO CAPACITY: Up to 13 troops/passengers

NOTES: First flight, 1972.

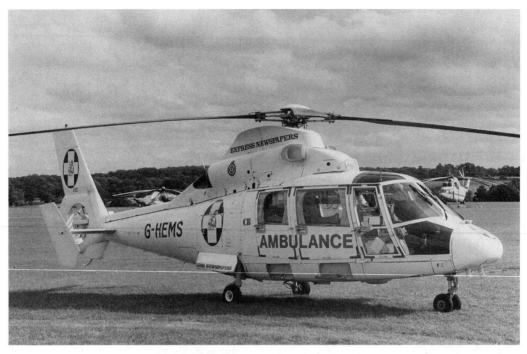

AEROSPATIALE SA 365 DAUPHINE II
Courtesy of Art-Tech/Aerospace/M.A.R.S/TRH/Navy Historical

COUNTRY OF ORIGIN: France

CREW: 1 or 2 pilots/USCG 2 pilots, flight mechanic, and a rescue swimmer

ROTOR DIAMETER: 39 ft. 2 in.

LENGTH: 37 ft. 6 in.

ARMAMENT: Armed variant; combinations of 7.62/12.7-mm machine guns, 20-mm cannon, and antitank missiles; ASW versions with radar, sonar, and torpedoes

POWERPLANT: Two 650-horsepower turboshafts; USCG two Lycoming LTS-101-750B turboshafts

AIRSPEED: 165 knots

RANGE: 464 nautical miles

CARGO CAPACITY: Up to 15 troops/passengers

NOTES: Introduced in March 1979, the SA 365 N was a major modification with extensive use of composites and retractable landing gear. SA 365F produced for antiship and SAR missions. U.S. Coast Guard uses SA 366, designated HH-65A Dolphin, with two Lycoming turboshafts. SA 365M Panther, a multirole military version, armed with 7.62/12.7-mm machine guns and HOT ATGMs. SA 365N produced under license in China. More than 500 of all types produced or ordered.

EUROCOPTER AS 665 TIGRE/TIGER

Courtesy of Art-Tech/Aerospace/M.A.R.S/TRH/Navy Historical

COUNTRY OF ORIGIN: France/Germany

CREW: 2 pilots

ROTOR DIAMETER: 42 ft. 8 in.

LENGTH: 51 ft. 10 in. (rotors turning)

ARMAMENT: 30-mm cannon in nose turret. Mistral or Stinger AAMs, Hellfire, HOT, or Trigat ATGMs, FFARs, or 12.7/20-mm cannon pods

POWERPLANT: Two MTU/R-R Turbomeca MTR 390 turboshafts

AIRSPEED: 175 knots

RANGE: 405 nautical miles

CARGO CAPACITY: N/A

NOTES: Franco-German prototype of multirole helicopter first flown in April 1991, with original requirement of 427, but only 170 ordered. Australia selected the Tiger for its military in 2001. First production models delivered to German Army in March 2002.

ATLAS AIRCRAFT/DENEL AH-2 ROOIVALK
Courtesy of Fullhouse Imaging/Denel Aviation

COUNTRY OF ORIGIN: South Africa

CREW: Pilot and copilot/gunner

ROTOR DIAMETER: 51 ft. 1 in.

LENGTH: 61 ft. 5 in.

ARMAMENT: 20-mm cannon, 70-mm FFARs, and ZT6 Mokopa ATGWs or Mistral AAM

POWERPLANT: 2 Topaz 2,000-horsepower (DENAL manufactured Turbomeca Makila 1A1) turboshafts

AIRSPEED: 167 knots

RANGE: 720 nautical miles

CARGO CAPACITY: N/A

NOTES: First flown in 1990, the Rooivalk initially was equipped with Puma transmissions and several other components supplied by Eurocopter. The SAAF received its first AH-2 in 1999.

AGUSTA A109
Courtesy of Art-Tech/Aerospace/M.A.R.S/TRH/Navy Historical

COUNTRY OF ORIGIN: Italy

CREW: 1 or 2 pilots

ROTOR DIAMETER: 36 ft. 1 in.

LENGTH: 37 ft. 6 in.

ARMAMENT: A109CM/HA, 7.62-mm or .50-caliber machine gun, TOW ATGW, 2.75-in. FFAR, .50-caliber gun pods

POWERPLANT: 2 Allison 250C 420-horsepower, or Turbomeca Arriel turboshafts

AIRSPEED: 168 knots

RANGE: 352 nautical miles

CARGO CAPACITY: 6 or 7 passengers

NOTES: Prototype first flown in 1971, first deliveries in 1976. More than 600 A109s sold worldwide to civil and military markets.

Available in both civilian and military variants with retractable landing gear. Military versions perform utility, electronic countermeasures, ambulance, scout/attack, and antitank roles. New A109K military model has fixed landing gear, a longer nose housing, advanced avionics, and fire control equipment.

AGUSTA A129 MANGUSTA (MONGOOSE)
Courtesy of Art-Tech/Aerospace/M.A.R.S/TRH/Navy Historical

COUNTRY OF ORIGIN: Italy
CREW: 2 pilots
ROTOR DIAMETER: 39 ft.
LENGTH: 46 ft. 10 in.
ARMAMENT: 20-mm cannon in nose turret; Stinger or Mistral AAM, TOW ATGW, 2.75-in. FFAR, .50-caliber or 20-mm cannon pods
POWERPLANT: Two R-RGEM 1004, LHTEC T800 turboshafts

AIRSPEED: 159 knots
RANGE: 303 nautical miles
CARGO CAPACITY: 2,204 pounds
NOTES: 1983 first flight, with deliveries in 1990. Multirole, ship-borne, and international versions produced. Approximately 60 A129s produced.

BELL MODEL 47; H-13 SIOUX
Courtesy of Art-Tech/Aerospace/M.A.R.S/TRH/Navy Historical

COUNTRY OF ORIGIN: USA

CREW: 1 pilot

ROTOR DIAMETER: 37 ft. 1 in.

LENGTH: 43 ft. 2 in.

ARMAMENT: Most models N/A. U.S. Army scout version in Vietnam, 2.30-caliber machine guns mounted on skids

POWERPLANT: 200-horsepower Franklin or 185-horsepower Lycoming piston engine

AIRSPEED: Cruise, 70 knots; maximum, 91 knots

RANGE: 217 nautical miles

CARGO CAPACITY: 1 to 2 passengers, or two litters mounted on skids for air ambulance version, or 500 pounds' payload

NOTES: Bell 47 resulted from Bell Ship One flown in 1945 and became first civilian certificated helicopter. First models ordered by U.S. Army and Navy. Built under license in Italy as Agusta Bell 47 and by Westland in the United Kingdom as the Sioux AH-1. All models, except 47-B3, have familiar bubble canopy and open tailboom.

BELL 204 UH-1 IROQUOIS
Courtesy of Art-Tech/Aerospace/M.A.R.S/TRH/Navy Historical

COUNTRY OF ORIGIN: USA

CREW: 2 pilots and a crewchief plus door gunner in combat

ROTOR DIAMETER: 48 ft.

LENGTH: 53 ft. 11 in.

ARMAMENT: UH-1 utility, two 7.62-mm machine guns; UH-1 armed versions, multiple 7.62-mm machine guns 2.75-in. rocket pods, and 40-mm grenade launcher

POWERPLANT: One T-53 Lycoming turboshaft

AIRSPEED: Cruise, 80 knots; maximum, 128 knots

RANGE: Maximum, 332 nautical miles

CARGO CAPACITY: 8 to 11 passengers, maximum slingload 4,000 pounds

NOTES: Derived from XH-40, which flew in 1956; spawned multiple variants of commercial and military types. Civil and military versions differ principally by internal equipment. Fuji Heavy Industries of Japan began licensed production of the Huey in 1962 and built 90 UH-1Bs and 50 UH-1Hs, mostly for the Japanese Self-Defense Forces, but a few sold to civil operators. Built under license by Bell-Agusta in Italy from 1961 to 1974, and fitted with Rolls Royce Gnome turboshaft. U.S. Army armed B model with SS-11 and TOW missiles as early antitank helicopter; TOW equipped Hueys destroyed several NVA tanks in later stages of Vietnam War.

BELL 205 UH-1H IROQUOIS

Courtesy of Art-Tech/Aerospace/M.A.R.S/TRH/Navy Historical

COUNTRY OF ORIGIN: USA
CREW: Military, 2 pilots with crewchief, and door gunner; civilian, 1 pilot
ROTOR DIAMETER: 48 ft.
LENGTH: 41 ft. 10 in.
ARMAMENT: Two .30-caliber doorguns
POWERPLANT: 1,250-horsepower T-53 Lycoming turboshaft
AIRSPEED: Cruise, 110 knots; maximum, 120 knots

RANGE: 300 nautical miles
CARGO CAPACITY: 4,000-pound slingload
NOTES: First flight 1961. Longer and with more cabin space than Bell 204. Civilian version fitted with up to 15 passenger seats. Built under license in Italy, Japan (HU-1H), Taiwan, and Turkey.

BELL 206 JET RANGER, OH-58 KIOWA
Courtesy of Art-Tech/Aerospace/M.A.R.S/TRH/Navy Historical

COUNTRY OF ORIGIN: USA

CREW: Pilot and observer

ROTOR DIAMETER: 35 ft.

LENGTH: 41 ft. 2 in.

ARMAMENT: Kiowa Warrior, combination of 7.62-mm minigun or .50-caliber machine gun and 2.75 rockets; TOW or Hellfire AT missiles

POWERPLANT: OH-58D, one Allison 375–420-horsepower turboshaft

AIRSPEED: 120 to 140 knots, depending on model

RANGE: 300 nautical miles

CARGO CAPACITY: 150 pounds on platform installed on rear seats; Long Ranger with double doors on port side carries more cargo or 2 stretchers internally in EMS version

NOTES: First flight 1962, with original model 206A remaining in production until 1972 with 660 units manufactured. In 1971, 206 B appeared with 400-horsepower Allison engine; 1,619 delivered. Current Jet Ranger III has 420-horsepower engine and enlarged vertical fin. Kiowa has slightly larger rotor and military standard avionics. Beginning in 1984, U.S. Army began converting all OH-58As into the four-bladed OH-58 D observation and Kiowa Warrior models with Mast Mounted Sight System. Bell also produces the military export version 406 Combat Scout.

BELL 209 AH-1G-F COBRA
Courtesy of Art-Tech/Aerospace/M.A.R.S/TRH/Navy Historical

COUNTRY OF ORIGIN: USA

CREW: Pilot and copilot/gunner

ROTOR DIAMETER: 48 ft.

LENGTH: 58 ft.

ARMAMENT: 7.62-mm miniguns, 40-mm grenade launcher, 2.75 rockets; newer versions, 20-mm cannon and TOW or Hellfire missiles

POWERPLANT: One T-53 or one T 400; Super Cobra two GE T700-401 turboshafts

AIRSPEED: Cruise, 120 to 140 knots; maximum, 222 knots

RANGE: 350 nautical miles

NOTES: First flight, 1965. Several single-engine variants produced, including the world's first true attack helicopter, the U.S. Army AH-1G. Improved versions were designed for antiarmor roles and armed with TOW missiles and, the later S models, with a 20-mm cannon. Twin-engined versions include the AH 1J and T Sea Cobra and AH-1W Super Cobra operated by the U.S. Marine Corps and the Israeli Defense Force. During the 1970s, Iran bought the Cobra for its army, but lack of spare parts grounded most of the fleet.

BELL 212/412

Courtesy of Art-Tech/Aerospace/M.A.R.S/TRH/Navy Historical

COUNTRY OF ORIGIN: USA

CREW: 1 or 2 pilots and crewchief

ROTOR DIAMETER: 48 ft. 2 in.

LENGTH: 57 ft. 3 in.

ARMAMENT: Two 7.62-mm machine guns; wire guided missiles; ASW version, two homing torpedoes or depth charges,

POWERPLANT: Two PT-6T 900-horsepower turboshafts

AIRSPEED: 128 knots

RANGE: 227 nautical miles

CARGO CAPACITY: Up to 15 passengers

NOTES: First flew in 1968. Military designation UH-1N used by U.S. Navy and Marine Corps. CH-135 Canadian armed forces designation. Agusta 212 produced in Italy; the 212 ASW produced for the Italian Navy, Turkey, Iran (until 1979), and several other countries. After 1988 all 212s produced in Canada. The four-bladed version is 412; latest version is 412 SP.

BELL 214
Courtesy of Art-Tech/Aerospace/M.A.R.S/TRH/Navy Historical

COUNTRY OF ORIGIN: USA

CREW: 1 or 2 pilots

ROTOR DIAMETER: 49 ft. 11 in.

LENGTH: 42 ft. 4 in.

ARMAMENT: N/A

POWERPLANT: 1 Lycoming T-55 or two PT-7 turboshafts; Super Transport version, two GE CT-7 turboshafts

AIRSPEED: Cruise, 140 knots

RANGE: 483 kilometers

CARGO CAPACITY: 4,000-pound external slingload; 7,000 pounds internal

NOTES: Developed from Bell 205, the 214 first flew in 1974 and initially produced in large numbers for Iran. The civil version, 214 B, carries 14 passengers, or may be used for crop spraying or fire fighting. Super Transport ST version has longer fuselage and accommodates 18 passengers. Manufactured in Italy under license by Agusta.

BELL 412/AB412
Courtesy of Art-Tech/Aerospace/M.A.R.S/TRH/Navy Historical

COUNTRY OF ORIGIN: USA
CREW: 1 or 2 pilots
ROTOR DIAMETER: 46 ft.
LENGTH: 56 ft. 2 in.
ARMAMENT: AB-412 Griffon, .50-caliber machine gun, Sea Skua ASM, TOW ATGM, 2.75 FFAR, or 25-mm cannon pod
POWERPLANT: Two PT-6 turboshafts

AIRSPEED: 122 knots
RANGE: 402 nautical miles
CARGO CAPACITY: Up to 14 passengers
NOTES: Essentially a 212 with a four-bladed rotor system. First flight in 1979. Production transferred to Canada in 1989; also licenses in Italy and Indonesia. More than 500 of all models produced.

BELL-BOEING V-22 OSPREY
Courtesy of Art-Tech/Aerospace/M.A.R.S/TRH/Navy Historical

COUNTRY OF ORIGIN: USA

CREW: 2 pilots and crewchief

ROTOR DIAMETER: 38 ft. 1 in.

LENGTH: 63 ft. (wings folded)

ARMAMENT: 7.62-mm or .50-caliber machine guns

POWERPLANT: Two Allison T406-AO-400 6,150-horsepower turboshafts

AIRSPEED: Cruise, 100 knots in helicopter mode; 315 knots in airplane mode

RANGE: 515 nautical miles

CARGO CAPACITY: 24 combat-equipped troops, or 12 litters; or 20,000 pounds of internal cargo, or 15,000-pound slingload

NOTES: Developed from XV-15 program, first prototypes flew in 1989. U.S. Marine Corps took first delivery in 1999 for test and evaluation.

BOEING 500 DEFENDER
(Hughes Aircraft and MacDonald/Douglas)

COUNTRY OF ORIGIN: USA

CREW: 1 or 2 pilots

ROTOR DIAMETER: 26 ft. 4 in.

LENGTH: 23 ft.

ARMAMENT: 7.62-mm minigun and 2.75 FFARs, .50-caliber machine gun, Stinger AAM, TOW or Hellfire missiles

POWERPLANT: 1 Allison 250 420-horsepower turboshaft

AIRSPEED: 120 to 140 knots, depending on version

RANGE: 215 nautical miles

CARGO CAPACITY: 1,000 pounds or 3 to 7 passengers, depending on model; 2,000-pound slingload

NOTES: Introduced as U.S. Army OH-6 Cayuse in 1963. More current versions have more powerful engines and updated avionics. The 500D and MD versions are distinguished by a T tail and five-bladed rotor system. The MD Defender is the military variant armed with a variety of machine guns and rockets. A newer 500 MG is equipped with a mast-mounted sighting system. U.S. Special Operations forces are equipped with both the Night Fox for night operations and the MH-6/8. The 500E has a longer and more streamlined nose, and the 500F Lifter is designed for high-temperature and high-altitude operations. Manufactured under license by Kawasaki in Japan and Breda-Nardi in Italy. Operated worldwide.

BOEING AH-64A AND D APACHE
(Hughes Aircraft and McDonnell-Douglas)

COUNTRY OF ORIGIN: USA

CREW: Pilot and copilot/gunner

ROTOR DIAMETER: 48 ft.

LENGTH: 58 ft. 3 in.

ARMAMENT: 30-mm chain gun, 2.75-in FFAR, and 16 Hellfire antitank missiles

POWERPLANT: Two T700-GE-701 1,698-horsepower, D T700-GE-701C 1,940-horsepower turboshafts

AIRSPEED: Maximum cruise, 150 knots; maximum, 200 knots

RANGE: 220 nautical miles without wingtanks

CARGO CAPACITY: N/A

NOTES: Designed as a tandem two-seat, all-weather attack helicopter, the AH-64 prototype lifted off in 1975. Upgrades led to the even more lethal AH-64D in 1997.

BOEING/VERTOL H46A-F SEA KNIGHT
Courtesy of Art-Tech/Aerospace/M.A.R.S/TRH/Navy Historical

COUNTRY OF ORIGIN: USA and Japan

CREW: 2 pilots, crewchief, and door gunner

ROTOR DIAMETER: 51 ft.

LENGTH: 44 ft. 10 in.

ARMAMENT: 7.62-mm or .50-caliber machine gun (door-mounted)

POWERPLANT: Two T58 turboshafts

AIRSPEED: 144 knots

RANGE: 206 nautical miles

CARGO CAPACITY: 25 troops

NOTES: First flown in 1958, adopted by U.S. Marines as ship- or land-based assault/utility helicopter. Canada operates Ch-113 and Sweden purchased Rolls-Royce–powered HKP4. Kawasaki produced aircraft under contract as 107, and military version Kv-107.

BOEING/VERTOL CH-47A-D, MH-47E CHINOOK
Courtesy of Art-Tech/Aerospace/M.A.R.S/TRH/Navy Historical

COUNTRY OF ORIGIN: USA

CREW: 2 pilots, crewchief, and door gunner

ROTOR DIAMETER: 60 ft.

LENGTH: Fuselage, 50 ft. 11 in.

ARMAMENT: 7.62-mm or .50-caliber doorguns

POWERPLANT: Two Lycoming T55-L-712 turboshafts

AIRSPEED: 154 knots

RANGE: 505 nautical miles

CARGO CAPACITY: 33 to 55 troops, or 20,000 pounds internally, or 28,000 pounds on external hooks

NOTES: Maiden flight 1961. All U.S. Army A-C models remanufactured to D configuration. Civil version is 234, international version is 414. RAF operates machine as HC Mk I. Produced under contract by Meribionali in Italy. More than 1,000 of all versions produced.

BOEING-SIKORSKY RAH-66 COMANCHE
Courtesy of Art-Tech/Aerospace/M.A.R.S/TRH/Navy Historical

COUNTRY OF ORIGIN: USA
CREW: 2 pilots
ROTOR DIAMETER: 40 ft.
LENGTH: 46 ft. 10 in.
ARMAMENT: 20-mm nose turret cannon, internal weapons bays; Stinger AAM, Hellfire ATGM, 2.75 FFAR, or cannon pods on wings

POWERPLANT: Two LHTEC-800 turboshafts
AIRSPEED: 175 knots
RANGE: 300 nautical miles
CARGO CAPACITY: N/A
NOTES: LHX stealth helicopter, YRAH-66 prototype flew in 1996. 1,200 aircraft planned for U.S. Army.

HILLER 360; UH-12; H-23 RAVEN
Courtesy of Art-Tech/Aerospace/M.A.R.S/TRH/Navy Historical

COUNTRY OF ORIGIN: USA

CREW: 1 pilot

ROTOR DIAMETER: 35 ft. 5 in.

LENGTH: 28 ft. 6 in.

ARMAMENT: N/A

POWERPLANT: Models A, B, and C 210-horsepower Franklin and model E 350 Lycoming piston engines

AIRSPEED: Maximum, 85 knots; cruise, 75 knots

RANGE: 225 miles

CARGO CAPACITY: 2 passengers, or 2 external litters

NOTES: Introduced in 1948 and modified with goldfish bowl cabin in 1950. Hiller Aviation reorganized in 1973 and acquired by Rogerson Aircraft in 1984, which resumed production of UH-12 in 1984.

ROGERSON (HILLER) RH-110
Courtesy of Art-Tech/Aerospace/M.A.R.S/TRH/Navy Historical

COUNTRY OF ORIGIN: USA
CREW: 1 pilot
ROTOR DIAMETER: 35 ft. 4 in.
LENGTH: 29 ft. 9 in.
ARMAMENT: N/A
POWERPLANT: One T-63 turboshaft
AIRSPEED: 90 knots
RANGE: 315 nautical miles

CARGO CAPACITY: Up to 3 passengers
NOTES: Evolved from the Hiller OH-5A and known originally as the Hiller 1100, first flew in 1963. Only about 250 of both military and civilian types produced. Some South American military forces operate the armed RH-1100 M Hornet.

HINDUSTAN AERONAUTICS LTD DHRUVS ALH
Courtesy of Art-Tech/Aerospace/M.A.R.S/TRH/Navy Historical

COUNTRY OF ORIGIN: India
CREW: 2 pilots
ROTOR DIAMETER: 43 ft. 4 in.
LENGTH: 52 ft. 1 in. (rotors turning)
ARMAMENT: Provisions for 20-mm cannon; probable later armed version with AAM, ATGM, FFAR; naval version depth charges, or torpedoes
POWERPLANT: 2 Turbomeca TM 333-2B turboshafts
AIRSPEED: 178 knots
RANGE: 432 nautical miles

CARGO CAPACITY: 12 to 14 troops; 3,300 pounds slingload
NOTES: With German assistance the first prototype flew in August 1992. Designed to meet the requirements of the Indian Air Force, Army, and Navy, the first seven production aircraft were delivered in April 2002. Naval variant equipped with retractable wheeled landing gear instead of skids.

HUGHES, SWEITZER 269, 300, TH-55A OSAGE
Courtesy of Art-Tech/Aerospace/M.A.R.S/TRH/Navy Historical

COUNTRY OF ORIGIN: USA
CREW: 1 pilot
ROTOR DIAMETER: 25 ft. 3 in.
LENGTH: 22 ft. 3 in.
ARMAMENT: N/A
POWERPLANT: One Lycoming piston engine; one turboshaft
AIRSPEED: 70 mph
RANGE: 204 statute miles
CARGO CAPACITY: 1 passenger

NOTES: Designed in 1955, the helicopter first flew the following year. By 1968, more than 1,200 had been produced for military training and civilian use. Sweitzer acquired production rights in 1983 and produced a turbine-engined version. Several U.S. police forces received surplus TH-55s.

KAMAN HH-43B HUSKY
Courtesy of Art-Tech/Aerospace/M.A.R.S/TRH/Navy Historical

COUNTRY OF ORIGIN: USA

CREW: 1 or 2 pilots and a rescue hoist operator

ROTOR DIAMETER: 47 ft. 9 in.

LENGTH: 25 ft. 7 in.

ARMAMENT: N/A

POWERPLANT: One 825-horsepower Lycoming T-53 gas turbine

AIRSPEED: Maximum 104 knots

RANGE: 350 statue miles

CARGO CAPACITY: Rescue 4 passengers, 8 troops on folding seats, or 4 litters and 1 medic hoist, 600 pounds

NOTES: A derivative of the K-225 and introduced in December 1958. USAF used the Husky for SAR and firefighting. The HH-43 claimed the best safety record of any SAR helicopter during the Vietnam War. Some 276 units, of all series, produced. Exported to Burma, Pakistan, Colombia, Morocco, and Thailand.

KAMAN H-2 SEA SPRITE
Courtesy of Art-Tech/Aerospace/M.A.R.S/TRH/Navy Historical

COUNTRY OF ORIGIN: USA
CREW: 2 pilots and equipment operators or parajumpers
ROTOR DIAMETER: 44 ft. 4 in.
LENGTH: 38 ft. 4 in.
ARMAMENT: 7.62-mm doorgun, 2 to 4 Penguin or Sea Skua ASMs, torpedoes, or Hellfire ATGM
POWERPLANT: Two T58 or GE T-700 turboshafts
AIRSPEED: 150 knots
RANGE: 400 nautical miles
CARGO CAPACITY: Single engine 3,500 pounds; twin 4,400 pounds
NOTES: First flown in 1959, the Sea Sprite is capable of antisubmarine, antimissile defense, search and rescue, and utility roles. In 1963, the single-engined UH-2A entered service with the U.S. Navy. Subsequent versions were twin-engined and include the HH-2D for coastal and geodetic survey, SH-2d LAMPS (antisubmarine and utility aircraft), and SH-2F improved LAMPS version. Most current SH-2G Super Sea Sprite began production in 1981, with more powerful engines and advanced avionics. Withdrawn from U.S. Navy service in 2001, but still operated by Australia, Egypt, New Zealand, and Poland.

KAMOV KA-25 HORMONE

Courtesy of Art-Tech/Aerospace/M.A.R.S/TRH/Navy Historical

COUNTRY OF ORIGIN: USSR

CREW: 2 pilots and 3 sensor operators

ROTOR DIAMETER: 51 ft. 7 in.

LENGTH: 32 ft.

ARMAMENT: Depth charges, torpedoes, and guided missiles

POWERPLANT: 2 Grunshenkov turbo-shafts

AIRSPEED: 119 knots

RANGE: 220 nautical miles

CARGO CAPACITY: Cargo version 12 passengers

NOTES: Standard Soviet/Russian ASW helicopter, distinguished by coaxial rotors and a large chin radome; B model configured to provide over-the-horizon targeting information for cruise missiles and naval gunfire; C model used for SAR. Ka-25K, without radome, designed for commercial crane and utility duties. By 1973 at least 170 of all variants produced; used in USSR/Russia, Syria, Vietnam, and India.

KAMOV KA-26 HOODLUM

Courtesy of Art-Tech/Aerospace/M.A.R.S/TRH/Navy Historical

COUNTRY OF ORIGIN: USSR

CREW: 1 or 2 pilots

ROTOR DIAMETER: 42 ft. 8 in.

LENGTH: 25 ft. 7 in.

ARMAMENT: N/A

POWERPLANT: 2 Vedeneyev M-14V-26 325-horsepower radial piston engines

AIRSPEED: Maximum 95 knots

RANGE: 220 nautical miles at maximum load

CARGO CAPACITY: 7 passengers or 1,980 pounds

NOTES: Prototype first flown in 1965. Designed as light cargo and agricultural helicopter. Interchangeable pods adapt machine as air ambulance, crop duster, or geodetical survey platform. Single-engine Ka-126 Hoodlum B produced in Romania. Exported worldwide.

KAMOV KA-27A/B HELIX
Courtesy of Art-Tech/Aerospace/M.A.R.S/TRH/Navy Historical

COUNTRY OF ORIGIN: USSR

CREW: 2 pilots and 2–3 sensor operators

ROTOR DIAMETER: 52 ft. 2 in.

LENGTH: 37 ft. 8 in.

ARMAMENT: 7.62-mm doorgun, dipping sonar, AT-6 Spiral ATGMs, depth charges, torpedoes, bombs, or 23/30-mm gun pod

POWERPLANT: 2 Klimov TV3-117V turboshafts

AIRSPEED: Maximum, 146 knots

RANGE: 432 nautical miles

CARGO CAPACITY: Helix-B15 troops

NOTES: Tested in 1973 and introduced into active service in 1981, similar to Ka-25, but smaller and lighter. Helix-A is ASW version, and B model (introduced in 1985) is utilized as naval infantry assault helicopter. Ka-27PS Helix-D modified for SAR. Ka-32S/T are exported. About 270 of all models produced; also utilized by Algeria, China, India, Laos, and Vietnam.

KAMOV KA-50/52 HOKUM
Courtesy of Art-Tech/Aerospace/M.A.R.S/TRH/Navy Historical

COUNTRY OF ORIGIN: USSR/Russia

CREW: 1 pilot

ROTOR DIAMETER: 47 ft. 7 in.

LENGTH: 52 ft. 6 in. (rotors turning)

ARMAMENT: 30-mm cannon in fuselage, AA-11 Archer AAM, AT-6 ATGM, 80- or 120-mm rocket pods, and/or bombs

POWERPLANT: 2 Klimov TV3-117 turboshafts

AIRSPEED: 210 knots

RANGE: Reported 590 nautical miles

CARGO CAPACITY: 6,600 pound armament

NOTES: Designed as a single-seat, close air support helicopter, the first prototype lifted off in June 1982. Equipped with an ejection seat, the Ka-50 was ordered into limited production, but only prototypes of the side-by-side Ka-52 were exhibited.

KAWASAKI OH-1
Courtesy of Art-Tech/Aerospace/M.A.R.S/TRH/Navy Historical

COUNTRY OF ORIGIN: Japan
CREW: 2 pilots
ROTOR DIAMETER: 38 ft. 1 in.
LENGTH: 39 ft. 4 in. (rotors turning)
ARMAMENT: Type 91 IR-guided AAM
POWERPLANT: Two Mitsubishi TS1-10QT turboshafts
AIRSPEED: 150 knots

RANGE: 300 nautical miles
CARGO CAPACITY: N/A
NOTES: Designed with a fenestron-type tailrotor and fixed wheel landing gear, the prototype scout/observation OH-1 first flew in August 1996. Japanese Self Defense Force has ordered approximately 200 to date.

MIL MI-1 HARE
Courtesy of Art-Tech/Aerospace/M.A.R.S/TRH/Navy Historical

COUNTRY OF ORIGIN: USSR
CREW: 1 pilot and 1 observer
ROTOR DIAMETER: 47 ft.
LENGTH: 39 ft. 4 in.
ARMAMENT: N/A
POWERPLANT: 1 Ivchenko AI-26V 575-horsepower radial engine
AIRSPEED: Maximum, 110 knots
RANGE: 320 nautical miles
CARGO CAPACITY: 3 passengers
NOTES: Developed from the EG-1, the Mi-1 became the first helicopter produced in any quantity in the Soviet Union. Several hundred civilian and military Mi-1s were produced in the USSR, and beginning in 1955, 1,700 in Poland, designated the WSK-Swidnik SM-1. Used in observation, liaison, rescue, air ambulance, and training missions; Albania, Afghanistan, Cuba, Czechoslovakia, Finland, Iraq, Poland, Syria, the United Arab Emirates, and Yemen procured variants of the Mi-1.

MIL MI-2 HOPLITE
Courtesy of Art-Tech/Aerospace/M.A.R.S/TRH/Navy Historical

COUNTRY OF ORIGIN: USSR/Poland

CREW: Pilot and 1 observer

ROTOR DIAMETER: 47 ft. 7 in.

LENGTH: 57 ft. 2 in.

ARMAMENT: Mi-2URP 2–4 AT-3 Sagger ATGM, or Strela 2 AAM

POWERPLANT: 2 Isotov 431-horsepower turboshaft

AIRSPEED: Maximum, 115 knots

RANGE: 230 nautical miles (internal fuel)

CARGO CAPACITY: Up to 8 passengers

NOTES: Developed from Mi-1 in 1961 and in 1964 WSK PZL in Poland assumed production and marketing. More than 5,450 produced in both civil and military variants. Mi-2s are used as air ambulances, search and rescue, agricultural work, and utility transports. Military versions sometimes carry rockets and wire-guided antitank missiles on pylons attached to the fuselage.

MIL MI 4 HOUND

Courtesy of Art-Tech/Aerospace/M.A.R.S/TRH/Navy Historical

COUNTRY OF ORIGIN: USSR

CREW: 1 or 2 pilots and a mechanic/gunner

ROTOR DIAMETER: 69 ft.

LENGTH: 55 ft. 2 in.

ARMAMENT: Variants with 12.7-mm machine guns and 57-mm rockets. Antisubmarine models are equipped with magnetic anomaly detection gear and a nose-mounted search radar.

POWERPLANT: 1 Svetzov 1,700-horsepower radial piston engine

AIRSPEED: Cruise, 85 knots

RANGE: 320 nautical miles

CARGO CAPACITY: 5,350 pounds

NOTES: First production models 1952, with more than 3,500 military and civilian models manufactured. Exported to all Soviet satellites, as well as Egypt, India, and Iraq. Soviet Army version capable of carrying up to 16 combat equipped troops. Civilian model Mi 4P carries 11 passengers. Either military or civilian versions can lift a 2,800-pound slingload, or other bulky cargo, which may be loaded through the rear clamshell doors. The agricultural/crop spraying version is the Mi 4. The People's Republic of China produced the Mi 4 under license.

MIL MI-6 HOOK
Courtesy of Art-Tech/Aerospace/M.A.R.S/TRH/Navy Historical

COUNTRY OF ORIGIN: USSR
CREW: 2 pilots, navigator, flight engineer, and radio operator
ROTOR DIAMETER: 114 ft. 10 in.
LENGTH: Fuselage 108 ft. 10 in., 137 overall
ARMAMENT: Some with DShK 12.7-mm nose-mounted machine gun
POWERPLANT: Two Soloviev D-25 V (TV-2BM) 5,000-horsepower turboshafts
AIRSPEED: Maximum, 160 knots
RANGE: 320 nautical miles, 540 nautical miles ferry
CARGO CAPACITY: 65 to 70 passengers, 90 combat troops, 26,400 internal, 17,650 pounds' external cargo

NOTES: First flown in June 1957 as world's largest helicopter; 860 produced, used mainly by Soviet/Russian military. Aeroflot operated the Mi-6P with square windows and no clamshell doors. Developed into Mi-10 Flying Crane and Mi-22C. Vehicles, artillery, and missiles loaded through rear clamshell doors, and heavy loads carried externally. Algeria, Egypt, Ethiopia, Bulgaria, Iraq, Syria, Peru, and Vietnam bought military versions.

MIL MI-8 /MI-17 HIP C-K
Courtesy of Art-Tech/Aerospace/M.A.R.S/TRH/Navy Historical

COUNTRY OF ORIGIN: USSR

CREW: Two pilots and flight engineer

ROTOR DIAMETER: 70 ft.

LENGTH: 61 ft.

ARMAMENT: Combinations of 12.7-mm machine gun in nose, SA-2 Swatter/AT-3 Sagger ATGMs, 57- and 80-mm rocket pods, 23-mm cannon, and bombs

POWERPLANT: Two Klimov TV2-117AG (Mi-8)/ TV3-117MT (Mi-17) turboshafts

AIRSPEED: 135 knots

RANGE: 200 nautical miles (545 with auxiliary fuel)

CARGO CAPACITY: Up to 32 combat troops, or 8,800 pounds' internal cargo, or 6,614 pounds slingload

NOTES: Designed to replace Mi-4, first flown in June 1961; used by Soviet and Russian forces and Aeroflot. Military versions denoted by round windows and armed with machine guns and 57-mm rockets. Later version designed and equipped for ECM operations. Introduced in August 1975, Mi-17 employed Mi-8 fuselage and Mi-14 engines; latest version with upgraded engines is Mi-17 Hip H. More than 10,000 of all variants manufactured and used by Armenia, Azerbaijan, Afghanistan, Algeria, Angola, Belarus, Bulgaria, Cambodia, the Commonwealth of Independent States, Croatia, Cuba, Czech Republic, Egypt, Germany, Guyana, Hungary, Iran, Iraq, Madagascar, Mongolia, Mozambique, Nicaragua, North Yemen, People's Republic of China, Slovakia, South Yemen, Sudan, Syria, Ukraine, Vietnam, Yugoslavia, and Zambia.

MIL MI-10/MI-10K HARKE
Courtesy of Art-Tech/Aerospace/M.A.R.S/TRH/Navy Historical

COUNTRY OF ORIGIN: USSR
CREW: 2 pilots and flight engineer
ROTOR DIAMETER: 114 ft. 10 in.
LENGTH: 107 ft. 9 in.
ARMAMENT: N/A
POWERPLANT: 2 Soloviev D-25 V (TV-2BM) 5,000-horsepower turboshafts
AIRSPEED: Cruise, 95 knots; maximum, 120 knots
RANGE: 135 nautical miles; 350 nautical miles with ferry tanks

CARGO CAPACITY: 36,300-pound slingloads and 28 passengers internally.
NOTES: Developed from Mi-6, the Mi-10 flying crane first flew in 1961. Utilized by both Soviet/Russian forces and Aeroflot. Introduced in 1966 the Mi-10K Harke B fuselage is modified with a shorter landing gear and a gondola under the nose. A closed-circuit TV system allows Mi-10 crews to monitor slingloads.

MIL MI-14 HAZE
Courtesy of Art-Tech/Aerospace/M.A.R.S/TRH/Navy Historical

COUNTRY OF ORIGIN: USSR

CREW: 2 pilots and 2 mission technicians

ROTOR DIAMETER: 69 ft. 10 in.

LENGTH: 59 ft. 6 in.

ARMAMENT: Torpedoes, depth charges, and nuclear depth bomb

POWERPLANT: Klimov TV3-117M turboshafts

AIRSPEED: Maximum, 125 knots

RANGE: 612 nautical miles

CARGO CAPACITY: 10 passengers

NOTES: Developed from Mi-8 in 1969, the Haze is a shore-based ASW helicopter. A boat-hulled fuselage, with a large radome under the nose, retractable landing gear, and a magnetic anomaly detector mounted to the underside of the tailboom differentiates the silhouette from that of the Mi-8. Three versions currently in service: Mi-14 PL Haze AASW, Mi-14BT Haze B mine countermeasures, and Mi-14PS Haze C Search and Rescue; 250 of all models produced. Also operated by Bulgaria, Cuba, Ethiopia, Libya, North Korea, Poland, and Syria.

MIL MI-24 HIND D/E/F/25/35
Courtesy of Art-Tech/Aerospace/M.A.R.S/TRH/Navy Historical

COUNTRY OF ORIGIN: USSR

CREW: Pilot, weapons operator, and flight engineer

ROTOR DIAMETER: 56 ft. 9 in.

LENGTH: 55 ft. 5 in.

ARMAMENT: 12.7-mm machine gun, 57-mm rockets, AT-6 Spiral ATGMs, AA-8 Aphid or AA-11 Archer AAMs, 23-mm gun pod; adapted to carry bombs as well.

POWERPLANT: 2 Usatov or Klimov TV3 turboshafts

AIRSPEED: 200 knots

RANGE: 175 nautical miles combat loaded

CARGO CAPACITY: 10 combat-equipped troops, 6,350 pounds of armament; 21,000 pounds combat loaded

NOTES: First flown in September 1969 and distinctive from the A model, which had side-by-side crew seating, the Hind has tandem seating for pilot and gunner and is heavily armed and armored; later versions have applique armor surrounding crew. Hind C is not equipped with nose gun, and F has wing-mounted gun pods and no nose turret. Specially equipped G models are used for radiation sampling. Tailrotor mounted on port side of vertical fin on later models. Hind 25 and 35 are export versions. More than 2,500 of all variants manufactured and used extensively by USSR in Afghanistan war. Over 2,100 Hinds sold to other countries.

MIL MI-26 HALO
Courtesy of Art-Tech/Aerospace/M.A.R.S/TRH/Navy Historical

COUNTRY OF ORIGIN: USSR
CREW: 2 pilots
ROTOR DIAMETER: 104 ft. 11 in.
LENGTH: 110 ft. 6 in.
ARMAMENT: N/A
POWERPLANT: 2 Lotorev turboshafts
AIRSPEED: Maximum, 160 knots

RANGE: 430 nautical miles
CARGO CAPACITY: 45,000 pounds internally, or 85 troops
NOTES: Heaviest helicopter produced in world, initially flown in 1977. Designed to operate in extreme winter conditions of Siberia.

MIL MI-28 HAVOC
Courtesy of Art-Tech/Aerospace/M.A.R.S/TRH/Navy Historical

COUNTRY OF ORIGIN: USSR/Russia
CREW: Pilot and copilot/gunner
ROTOR DIAMETER: 56 ft. 8 in.
LENGTH: 55 ft. 10 in.
ARMAMENT: 30-mm cannon under nose, AT-6 Spiral ATGMs, 80- or 120-mm rockets, 23-mm cannon pods, mine dispensers
POWERPLANT: 2 Klimov TV3-117VMA turboshafts
AIRSPEED: 162 knots

RANGE: 232 nautical miles
CARGO CAPACITY: Maximum gross weight, combat loaded, 25,000 pounds
NOTES: The Apache look-alike first flew in November 1982, brought to production stage, but no orders received for original or improved Mi-28N Night Hunter versions. Russian Army apparently selected Ka-50 Hokum over Mi-28.

MBB BO105

Courtesy of Art-Tech/Aerospace/M.A.R.S/TRH/Navy Historical

COUNTRY OF ORIGIN: Germany
CREW: 1 or 2 pilots
ROTOR DIAMETER: 32 ft. 3 in.
LENGTH: 38 ft.
ARMAMENT: 6 TOW or HOT missiles
POWERPLANT: 2 Allison 250 420-horse-power turboshafts
AIRSPEED: 130 knots
RANGE: 300 nautical miles
CARGO CAPACITY: 3 combat-loaded troops or 2 litters
NOTES: First flight February 1967, with several improved variants, all externally similar. BO 105 CBS has longer fuselage and increased seating capacity. The BO 105 M, or VBH, and PAH-1 are the most common military variants used in both scout/observation and antitank roles. The BO 105 is used extensively as an EMS helicopter in several countries. Built under license by Dirgantara in Indonesia and KAI in South Korea.

MBB/KAWASAKI BK-117
Courtesy of Art-Tech/Aerospace/M.A.R.S/TRH/Navy Historical

COUNTRY OF ORIGIN: Germany/Japan
CREW: 1 pilot
ROTOR DIAMETER: 36 ft.
LENGTH: 32 ft. 9 in.
ARMAMENT: N/A
POWERPLANT: 2 Lycoming 101 650- or 700-horsepower turboshafts
AIRSPEED: 145 knots

RANGE: 300 nautical miles
CARGO CAPACITY: EMS version 2 litters and two flight nurses/paramedics
NOTES: Prototype flew in 1979. Seats 7 passengers in executive version and up to 11 in high density forum. Produced in Germany, Japan, and Indonesia. Over 170 BK 117s delivered.

NH INDUSTRIES NH-90
Courtesy of Art-Tech/Aerospace/M.A.R.S/TRH/Navy Historical

COUNTRY OF ORIGIN: France/Germany/ Italy/Netherlands/Portugal

CREW: 2 pilots and 2 to 3 mission specialists

ROTOR DIAMETER: 53 ft. 5 in.

LENGTH: 64 ft. 2 in. (rotors turning)

ARMAMENT: Two 7.62-mm doorguns, Mk II ASMs, torpedoes, and depth charges

POWERPLANT: Two RTM 322 or GE T700-T6E turboshafts

AIRSPEED: 157 knots

RANGE: 650 nautical miles

CARGO CAPACITY: 14 to 20 troops; 10,400 pounds

NOTES: Designed as NATO's medium lift and naval helicopter for the twenty-first century, the NH-90 prototype took flight in December 1995. Of 657 machines required by the five collaborative countries, 305 have been ordered, including orders from Finland, Norway, and Sweden.

SIKORSKY R-4, HNS-1
Courtesy of Art-Tech/Aerospace/M.A.R.S/TRH/Navy Historical

COUNTRY OF ORIGIN: USA
CREW: 1 pilot and 1 observer
ROTOR DIAMETER: 18 ft.
LENGTH: 38 ft.
ARMAMENT: N/A
POWERPLANT: One 165-horsepower Warner piston engine
AIRSPEED: 70 mph

RANGE: Unknown
CARGO CAPACITY: N/A
NOTES: First U.S. helicopter used in combat operations; ordered by U.S. Army, Navy, USCG, and RAF during WW II. Sikorsky and Nash-Kelvinator built a total of 131 R-4s.

SIKORSKY S-51, R-5, H-5, HO2S-1, HO3S-1, HO3S-2
Courtesy of Art-Tech/Aerospace/M.A.R.S/TRH/Navy Historical

COUNTRY OF ORIGIN: USA

CREW: 1 pilot and 1 hoist operator, or 3 passengers

ROTOR DIAMETER: 33 ft.

LENGTH: 40 ft.

ARMAMENT: N/A

POWERPLANT: 165-horsepower Franklin piston engine, later models 245-horsepower engine

AIRSPEED: 125 mph

RANGE: Maximum, 250 miles

CARGO CAPACITY: 3 passengers or 1,100 pounds

NOTES: Although Sikorsky received a contract for 450 R-5s, the end of World War II reduced actual production to 65 units. In February 1946, Sikorsky began concentrating on a commercial version of the R-5, known as the S-51, and eventually manufactured 220 of this type. Westland, Ltd., built the WS-51 Dragonfly under license, producing 371 machines.

SIKORSKY R-6
Courtesy of Art-Tech/Aerospace/M.A.R.S/TRH/Navy Historical

COUNTRY OF ORIGIN: USA
CREW: 1 pilot and 1 hoist operator
ROTOR DIAMETER: 38 ft.
LENGTH: Unknown
ARMAMENT: N/A
POWERPLANT: 245-horsepower Franklin engine

AIRSPEED: 100 mph
RANGE: 387 miles
CARGO CAPACITY: 1,100 pounds
NOTES: Sikorsky and Nash-Kelvinator built a total of 229 R-6s.

SIKORSKY S 55, S 55T, H-19 CHICKASAW
Courtesy of Art-Tech/Aerospace/M.A.R.S/TRH/Navy Historical

COUNTRY OF ORIGIN: USA

CREW: 2 pilots and crewchief

ROTOR DIAMETER: 53 ft.

LENGTH: 42 ft.

ARMAMENT: N/A

POWERPLANT: 1 Wright R1300 700-horsepower piston engine; U.S. company Hemitech produced the turbine-powered S 55T

AIRSPEED: Cruise, 75 knots

RANGE: 250+ miles

CARGO CAPACITY: 9–12 passengers or 2,600 pounds

NOTES: First flight in 1949, 1,281 produced, and operated by civil and military services worldwide. Westland produced 477 machines, known as the Whirlwind. Mitsubishi and SNCAN also built more than 80 aircraft under license in Japan and France.

SIKORSKY H 58, S 58T, CH 34A-D, UH 34D CHOCTAW, VH-34D, UH 34E SEAHORSE
Courtesy of Art-Tech/Aerospace/M.A.R.S/TRH/Navy Historical

COUNTRY OF ORIGIN: USA

CREW: 2 pilots and crewchief

ROTOR DIAMETER: 56 ft. 1 in.

LENGTH: 47 ft. 2 in.

ARMAMENT: Some versions with 7.62-mm machine guns and rockets; French modified 1 variant with a 20-mm cannon

POWERPLANT: 1 Wright R1820 piston engine, or two PT 6 turboshafts

AIRSPEED: 114 knots maximum

RANGE: 300 nautical miles

CARGO CAPACITY: 18 troops or 2,700 pounds of internal cargo

NOTES: Total production 1,821 of military and civilian models. SH antisubmarine versions 34G and J are known as the Seabots; 58T, powered with two PT 6 turboshaft engines, offered new or as a conversion in 1970.

SIKORSKY S-61R CH-3E, HH3F PELICAN
Courtesy of Art-Tech/Aerospace/M.A.R.S/TRH/Navy Historical

COUNTRY OF ORIGIN: USA

CREW: 2 pilots and crewchief

ROTOR DIAMETER: 21 ft.

LENGTH: 57 ft. 3 in.

ARMAMENT: 7.62-mm doorguns

POWERPLANT: Two T58 turboshafts

AIRSPEED: 141 knots

RANGE: 400 nautical miles

CARGO CAPACITY: 30 troops with fitted hoist

NOTES: Developed from S-61 Sea King, first flew in June 1963, the Pelican has an in-flight refueling probe on starboard side, a rear loading ramp, and retractable landing gear. The Jolly Green Giant entered USAF service and was used for assault, transport, and search and rescue. USCG HH-3F equipped with nose radome. Over 100 aircraft produced, and Augusta still manufactures a version for the Italian Air Force under contract.

SIKORSKY S-61 H-3 SEA KING
Courtesy of Art-Tech/Aerospace/M.A.R.S/TRH/Navy Historical

COUNTRY OF ORIGIN: USA

CREW: 2 pilots and 2 system operators or parajumpers

ROTOR DIAMETER: 62 ft.

LENGTH: 72 ft. 8 in. (rotors turning)

ARMAMENT: 7.62-mm door-mounted miniguns, MK II or Exocet ASM, torpedoes, or depth charges

POWERPLANT: Two GE T58GE-10 turbo-shafts

AIRSPEED: 144 knots

RANGE: 542 nautical miles

CARGO CAPACITY: 28 troops or 8,000 pounds

NOTES: Initially designated HSS-2 by the U.S. military, the S-61 prototype first flew in March 1959. The boat-hulled amphibious ASW/SAR helicopter entered U.S. Navy service in September 1961, redesignated SH-3. Manufactured under license in Italy by Agusta, in Japan by Mitsubishi, and in the United Kingdom by Westland, 1,030 versions of all types were produced.

SIKORSKY S 61L-N

Courtesy of Art-Tech/Aerospace/M.A.R.S/TRH/Navy Historical

COUNTRY OF ORIGIN: USA
CREW: 1 or 2 pilots
ROTOR DIAMETER: 62 ft.
LENGTH: 72 ft. 10 in.
ARMAMENT: N/A
POWERPLANT: Two CT58 turboshafts
AIRSPEED: Cruise, 140 knots

RANGE: 430 nautical miles
CARGO CAPACITY: 28 passengers
NOTES: First flown in 1960, the S 61 is widely used as an airliner; the S 61 seats up to 28 passengers. S 61 N is amphibious. Agusta now holds the rights to produce the S 61.

SIKORSKY S 62, HH 52A

Courtesy of Art-Tech/Aerospace/M.A.R.S/TRH/Navy Historical

COUNTRY OF ORIGIN: USA

CREW: 2 pilots and 2 to 4 crewmembers

ROTOR DIAMETER: 53 ft.

LENGTH: 44 ft. 6 in.

ARMAMENT: N/A

POWERPLANT: One T58 turboshaft

AIRSPEED: Cruise, 85 knots

RANGE: 415 nautical miles

CARGO CAPACITY: 12 passengers or 6 litters

NOTES: First amphibious helicopter built by Sikorsky; first flight in 1957; built in both civil and military versions. U.S. Coast Guard bought 99 with the designation of HH 52A. Built under license by Mitsubishi in Japan.

SIKORSKY HH 3
Courtesy of Art-Tech/Aerospace/M.A.R.S/TRH/Navy Historical

COUNTRY OF ORIGIN: USA
CREW: 2 pilots and 2 to 4 crewmembers
ROTOR DIAMETER: 62 ft.
LENGTH: 54 ft. 9 in.
ARMAMENT: 7.62-mm machine guns
POWERPLANT: Two T58 turboshafts
AIRSPEED: Cruise, 120 knots
RANGE: 545 nautical miles
CARGO CAPACITY: 6,000-pound external load; up to 28 passengers
NOTES: First flown in 1959, the HH 3 is a multipurpose aircraft fitted for ASW, SAR, and a utility transport capable of accommodating 26 combat loaded troops. ASW version is designated SH-3A-J Sea King and equipped with radar and homing torpedoes. Known as CH-124 in the Canadian armed forces, HH 3 is operated in numerous countries and built under license in Italy by Agusta.

SIKORSKY S-64 E/F SKY CRANE, CH-54A/B TARHEE
Courtesy of Art-Tech/Aerospace/M.A.R.S/TRH/Navy Historical

COUNTRY OF ORIGIN: USA

CREW: 3 pilots and loadmaster

ROTOR DIAMETER: 71 ft. 10 in.

LENGTH: 70 ft. 3 in.

ARMAMENT: None

POWERPLANT: 2 Pratt & Whitney JTFD12A-5A 4,800-horsepower turboshafts

AIRSPEED: Cruise, 110 knots

RANGE: 200 nautical miles

CARGO CAPACITY: 20,000-pound sling-load; 45 troops, or 24 stretchers in detachable pods

NOTES: The Flying Crane took to the air in 1962, adopted by U.S. Army as CH-54 Tarhee heavy lift helicopter. Attachable pods were also equipped as field hospitals or mobile command posts.

SIKORSKY S-65A CH-53A/D HH-53 B/C SEA STALLION; CH-53E SUPER SEA STALLION
Courtesy of Art-Tech/Aerospace/M.A.R.S/TRH/Navy Historical

COUNTRY OF ORIGIN: USA

CREW: 2 pilots with variations of crew-chief, doorgunners, technical operators, and parajumpers

ROTOR DIAMETER: 72 ft. 3 in.; Ch-53E 78 ft. 9 in.

LENGTH: 67 ft. 2 in.; CH-53E 73 ft. 5 in.

ARMAMENT: Three 7.62-mm or .50-caliber doorguns

POWERPLANT: Two GE T64; CH-53E, three T64C turboshafts

AIRSPEED: 170 knots

RANGE: 468 nautical miles without in-flight refueling

CARGO CAPACITY: Up to 55 troops, or vehicles and guns loaded through rear cargo doors

NOTES: A heavy lift design first flew in October 1964; entered service with USMC in 1966 and adopted by USAF to replace the SAR version of H-3. RH-53D and MH-53E Sea Dragon operated by U.S. Navy in mine countermeasure operations; first flown in March 1974. MH-53J PAVE LOW is used by U.S. special operations for deep penetration missions. CH-53G built under license in Germany by VFW-Fokker (now part of Eurocopter). Currently 522 of all versions produced.

SIKORSKY S-70 UH-60, HH-60 BLACKHAWK, SH-60B SEAHAWK
Courtesy of Art-Tech/Aerospace/M.A.R.S/TRH/Navy Historical

COUNTRY OF ORIGIN: USA

CREW: 2 pilots and crewchief with doorgunner or technical operators, or parajumper

ROTOR DIAMETER: 53 ft. 3 in.

LENGTH: 50 ft.

ARMAMENT: Two 7.62-mm machine guns on utility version: combinations of 20/30-mm cannon, 2.75-inch rockets, and Hellfire ATGM, Stinger AAM, or mine dispensers. Naval versions mount MK 46 ASW torpedoes.

POWERPLANT: Two GE T-700 turboshafts

AIRSPEED: 194 knots

RANGE: 319 nautical miles, or 1,200 nautical miles with external tanks

CARGO CAPACITY: 12 combat loaded troops or litters. Maximum external load of 9,000 pounds

NOTES: Prototype flew in October 1974 in competition for U.S. Army Utility Tactical Transport Aircraft System and entered regular service in 1979. HH-60 Nighthawk SAR version produced for USAF. EH-60 Quick Fix is ECM version. U.S. Navy Light Airborne Multipurpose System (LAMPS) Seahawk tailboom folds for shipboard operations, and is equipped for ASW and ship surveillance/targeting. Westland builds the S-70 in United Kingdom as WS-70. Military and commercial versions sold worldwide. Well over 2,000 of all versions produced.

SIKORSKY S-76 SPIRIT
Courtesy of Art-Tech/Aerospace/M.A.R.S/TRH/Navy Historical

COUNTRY OF ORIGIN: USA

CREW: 2 pilots with crewchief and door gunner

ROTOR DIAMETER: 43 ft. 11 in.

LENGTH: 44 ft.

ARMAMENT: Combination of rockets, missiles, and machine guns on AUH-76

POWERPLANT: 2 Allison 250 or PT-6B turboshafts

AIRSPEED: 150 knots—8,380 pounds gross weight

RANGE: With 8 passengers, auxiliary fuel, 600 nautical miles

CARGO CAPACITY: Commercial, 14 passengers

NOTES: First flight in 1977, classified as all-weather helicopter. Produced in utility, armed, naval, and commercial variants.

WESTLAND SCOUT AND WASP
Courtesy of Art-Tech/Aerospace/M.A.R.S/TRH/Navy Historical

COUNTRY OF ORIGIN: United Kingdom
CREW: 1 pilot
ROTOR DIAMETER: 32 ft. 3 in.
LENGTH: 30 ft. 3 in.
ARMAMENT: N/A
POWERPLANT: Latest versions, one Rolls Royce Nimbus 710-horsepower turboshaft
AIRSPEED: Cruise, Scout 115 knots; Wasp 95 knots

RANGE: 270 nautical miles
CARGO CAPACITY: N/A
NOTES: First flew in 1958. British Army operated 150 Scouts with fixed skid gear. Navy Wasp version with wheeled landing gear first delivered in 1963. Versions still used in Brazil, New Zealand, Indonesia, Malaysia, and South Africa.

WESTLAND WESSEX
Courtesy of Art-Tech/Aerospace/M.A.R.S/TRH/Navy Historical

COUNTRY OF ORIGIN: United Kingdom

CREW: 2 pilots, or 1 pilot and 3 ASW operators

ROTOR DIAMETER: 56 ft.

LENGTH: 48 ft. 4 in.

ARMAMENT: Homing torpedoes; some Army variants with machine guns

POWERPLANT: 2 Gnome turboshafts

AIRSPEED: 100 knots

RANGE: 300 nautical miles

CARGO CAPACITY: 16 commandos or 8 stretchers and a medical attendant or 3,600-pound slingload

NOTES: First flown in 1958, used as general purpose and antisubmarine aircraft. Capable of carrying 10 passengers, the Wessex is employed worldwide by RAF, RN, and Royal Australian Navy in various configurations, including the HC 2 ambulance and HU5 commando assault transport.

WESTLAND SEA KING/COMMANDO
Courtesy of Art-Tech/Aerospace/M.A.R.S/TRH/Navy Historical

COUNTRY OF ORIGIN: United Kingdom

CREW: 2 pilots

ROTOR DIAMETER: 62 ft.

LENGTH: 55 ft. 9 in.

ARMAMENT: Homing torpedoes, depth charges

POWERPLANT: 2 Gnome turboshafts

AIRSPEED: 144 knots

RANGE: 540 nautical miles

CARGO CAPACITY: 21 royal guards

NOTES: An RN ASW helicopter, developed by Westland under license from S 61D, the HAR 3 is employed by the RAF for SAR. The Commando transport version seats 30 troops but has no amphibious capability. The Sea King HAS Mk6 is the latest RN version. India bought several of these helicopters for its armed services.

WESTLAND LYNX
Courtesy of Art-Tech/Aerospace/M.A.R.S/TRH/Navy Historical

COUNTRY OF ORIGIN: United Kingdom and France

CREW: 2 pilots, or pilot and gunners

ROTOR DIAMETER: 42 ft.

LENGTH: 39 ft. 6 in.

ARMAMENT: Army version: door-mounted machine guns, two 20-mm cannon, TOW HOT, or Hellfire AT-GMs; Navy: Sea Skua missiles, dipping sonar, Mk 44/46 or Sting Ray ASW torpedoes, or depth charges

POWERPLANT: 2 Rolls Royce Gem turboshafts

AIRSPEED: 145 (Super Lynx), 215 knots

RANGE: 320 nautical miles

CARGO CAPACITY: 10, 200 pounds

NOTES: 1 of 3 helicopters covered in a 1967 Anglo-French agreement, the Lynx first took to the air in March 1971, with two main versions, one for the British Navy and the other for the British Army. The AH-1 attack version entered service with the British Army in 1977; other Army versions are the AH-5, 7, and 9. The Royal Navy took first delivery in 1977; naval variants are: Lynx AHS-2, -3, and AHS-8. Several naval variants serve with the French Navy, Royal Netherlands Navy, German Navy, Argentine Navy, and Brazilian Navy. Qatar, Norway, and Denmark all operate some version of the Lynx. All naval versions and the AH-9 have wheeled landing gear, while the Army versions have skids. The Super Lynx, introduced in 1995, features revolutionary BERP main rotor blades, more powerful LHTEC CTS-800-4N engines, and advanced avionics and targeting systems.

WESTLAND 30

Courtesy of Art-Tech/Aerospace/M.A.R.S/TRH/Navy Historical

COUNTRY OF ORIGIN: United Kingdom
CREW: 1 or 2 pilots and crewchief
ROTOR DIAMETER: 43 ft. 7 in.
LENGTH: 47 ft.
ARMAMENT: N/A
POWERPLANT: Two GEM turboshafts
AIRSPEED: Cruise, 130 knots

RANGE: 370 nautical miles
CARGO CAPACITY: 22 combat troops
NOTES: Combines the Lynx rotor system and engines with a transport fuselage; first flew in 1979. The military version can accommodate up to 22 passengers.

GLOSSARY

Advancing Blade   The rotor blade experiencing an increase in relative wind because of airspeed.

Airfoil   A surfaced body designed to produce a force when subjected to an airflow.

Angle of Attack   The acute angle between the chord line of an airfoil and the resultant relative wind.

Angle of Incidence, also **Pitch Angle**   The angle of the rotor blade chord line with the plane of rotational tip path plane of the rotor system.

Antitorque   Method used to counteract the torque reaction that results from a turning rotor system.

Articulated Rotor System   A rotor system in which the hub is mounted rigidly to the mast and the individual blades are mounted on hinge pins, allowing them to flap up and down and move forward and backward. Individual blades feather by rotating about the blade grip retainer bearings.

Autorotation   The action of turning a rotor system by airflow, not by engine power. Airflow may be induced by forward movement or descending through the air.

Axis of Rotation   The center of rotation perpendicular to the plane of rotation.

Blowback   The tendency for the rotor disk to tilt aft as a result of flapping caused by the combined effects of dissymmetry of lift and transverse flow.

Camber   The curvature of an airfoil.

Center of Gravity   The point within an aircraft through which, for balance purposes, the total force of gravity is considered to act.

Center of Pressure   The point along the chord line of an airfoil through which all the aerodynamic forces are considered to act.

Chord   A longitudinal dimension of an airfoil section measured from the leading to the trailing edge.

Chord Line   A straight line connecting the leading and trailing edges of an airfoil.

Collective Feathering   The simultaneous change of pitch of all rotor blades in a rotor system in equal amounts.

Coning Angle   The angle between the plane of rotation and the rotor blade.

Cyclic Feathering   The change of pitch of individual rotor blades independently of the other rotor blades in the system.

Dissymmetry of Lift   The difference in lift exerted between the advancing half of the rotor disk and the retreating half.

Flapping   Up and down movement of a rotor blade.

Ground Effect   A condition of improved aircraft performance when operating near a surface; usually occurs in rotorcraft up to the radius of the rotor disk.

Induced Velocity (also known as Downwash or Rotorwash)   The induced vertical component of the relative wind.

Lead and Lag   Lead: movement of the rotor blade forward. Lag: movement of the rotor blade aft. Both are measured from the radial line through the center of the main rotor driveshaft.

Relative Wind   The airflow relative to an airfoil.

Retreating Blade   The rotor blade experiencing a decrease in relative wind caused by airspeed.

Retreating Blade Stall   A stall that begins at or near the tip of the blade, caused by the high angles of attack required to compensate for dissymmetry of lift.

Rigid Rotor System   A rotor system in which the rotor blades are fixed rigidly to the hub and not allowed to flap and lead and lag. The only action allowed is pitch change.

Rotational Velocity   The component of the relative wind produced by rotation of the rotor blade.

Semirigid Rotor System   A rotor system in which the blades are connected to the mast by a trunnion that allows blades to flap. Pitch change or feathering are allowed at the hub by the blade grip retainer bearing.

Settling with Power   A condition of powered flight whereby the helicopter settles through disturbed air caused by the helicopter's own downwash.

Stall   The condition of an airfoil in which it is at an angle of attack too great to produce lift.

Tailrotor   The antitorque device of a single rotor helicopter. Conventional control of this rotor is through the pilot's directional control pedals.

Tandem Rotor System   A main lifting rotor is used at either end of the helicopter. The rotors turn in opposite directions (counter-rotating), to counteract torque.

Tip Path Plane   A plane defined by the circle scribed by the average flight path of the blade tips in a rotor system; sometimes called the rotor disk.

Torque Effect   The reaction to the turning of the rotor system. In single rotor helicopters the fuselage tends to turn in the opposite direction of the rotor system.

Translational Flight   Any horizontal movement of the helicopter with respect to the air.

Translating Tendency   The tendency of a hovering single rotored helicopter to drift over the ground in the direction of tailrotor thrust.

Translational Lift   That point in flight in which the airflow through a rotor system changes from vertical to horizontal. Translational lift results in increased lift.

Adapted from the U.S. Army Field Manual 1-51, "Rotary Wing Flight."

BIBLIOGRAPHY

BOOKS

Adcock, Al. *H-3 Sea King in Action.* Carrollton, TX: Squadron/Signal Publications, 1995.

Ahnstrom, D. N. *The Complete Book of Helicopters.* New York: World Publishing Company, 1971.

Aircraft of the World: The Complete Guide. N.p./USA: International Masters Publishers AB, licensed to IMP, 1996.

Anderton, David, and Jay Miller. *Boeing Helicopter: The CH-47.* Arlington, TX: Aerofax, 1989.

Apostolo, Giorgio. *The Illustrated Encyclopedia of Helicopters.* New York: Bonanza Books, 1984.

Bilstein, Roger E. *Flight in America.* Rev. ed. Baltimore, MD: Johns Hopkins University Press, 1994.

Bowden, Mark. *Black Hawk Down: A Story of Modern War.* New York: Atlantic Monthly Press, 1999.

Brehm, Jack, and Pete Nelson. *That Others May Live: The True Story of a PJ, a Member of America's Most Daring Rescue Force.* New York: Crown Publishers, 2000.

Brooks, Peter W. *Cierva Autogiros: The Development of Rotary-Wing Flight.* Washington, DC: Smithsonian Institution Press, 1988.

Brown, David A. *The Bell Helicopter Textron Story: Changing the Way the World Flies.* Arlington, TX: Aerofax, 1995.

Carey, Keith. *The Helicopter.* Blue Ridge Summit, PA: Tab Books, 1986.

Chant, Christopher. *Fighting Helicopters of the 20th Century.* Christchurch, Dorset, England: Graham Beehag Books, 1996.

Coleman, J. D. *Pleiku: The Dawn of Helicopter Warfare in Vietnam.* New York: St. Martin's Press, 1988.

Cook, John L. *Rescue under Fire.* Atglen, PA: Schiffer Books, 1998.

Cowin, Hugh W. *Military Helicopters.* New York: Gallery Books, imprint of W. H. Smith Publishers, 1984.

Delear, Frank J. *Igor Sikorsky: His Three Careers in Aviation.* New York: Dodd, Mead, & Company, 1969.

Donald, David, ed. *The Complete Encyclopedia of World Aircraft*. New York: Barnes and Noble Books, 1997.

Dorr, Robert F., and Chris Bishop. *Vietnam Air War Debrief*. London: Aerospace Publishing, 1996.

Dowling, John. *RAF Helicopters: The First 20 Years*. London: Her Majesty's Stationery Office, 1992.

Ean, Nicholas, comp. and ed. *US Army Field Manual 1–51: Rotary Wing Flight*. Renton, WA: Aviation Supplies and Academics, 1988.

Endres, Gunter, and Michael J. Gething, comps. and eds. *Jane's Aircraft Recognition Guide*. 3d ed. New York: HarperCollins Publishers, 2002.

Everett-Heath, John. *Soviet Helicopters*. London: Jane's Publishing Company, 1983.

Francillon, Rene J. *Vietnam: The War in the Air*. New York: Arch Cape Press, 1987.

Fredriksen, John C. *Warbirds: An Illustrated Guide to US Military Aircraft, 1915–2000*. Santa Barbara, CA: ABC-CLIO, 1999.

Gablehouse, Charles. *Helicopters and Autogiros: A History of Rotating-wing and V/STOL Aviation*. Philadelphia: J. B. Lippincott Company, 1969.

Galvin, John R. *Air Assault: The Development of Airmobile Warfare*. New York: Hawthorn Books, 1969.

Gregory, Barry. *Vietnam Helicopter Handbook*. Wellingborough, England: Patrick Stephens Ltd., 1988.

Gregory, H. F. *Anything a Horse Can Do: The Story of the Helicopter*. Cornwall, NY: Cornwall Press, 1944.

Gunston, Bill, ed. *The Encyclopedia of Modern Warplanes*. New York: Barnes & Noble Books, 1995.

Gunston, Bill, and Mike Spick. *Modern Fighting Helicopters*. London: Crescent Books, 1996.

Gurney, Gene. *Vietnam: The War in the Air*. New York: Crown Publishers, 1985.

Harding, Stephen. *US Army Aircraft since 1947*. Stillwater, MN: Specialty Press, 1990.

Heatley, Michael. *The Illustrated History of Helicopters*. New York: Bison Books, 1985.

Higham, Robin, John T. Greenwood, and Von Hardesty. *Russian Aviation and Air Power in the Twentieth Century*. London: Frank Cass, 1998.

Hirschberg, Michael, and David K. Daley. *US and Russian Helicopter Development in the 20th Century*. N.p.: American Helicopter Society International, 2000.

Howze, Hamilton H. *A Cavalryman's Story: Memoirs of a Twentieth-Century Army General*. Washington, DC: Smithsonian Institution Press, 1996.

Hunt, William E. *Helicopter: Pioneering with Igor Sikorsky*. Shrewsbury, England: Airlife Publishing, 1998.

Johnson, Lawrence H., III. *Winged Sabers: The Air Cavalry in Vietnam, 1965–1973*. Harrisburg, PA: Stackpole Books, 1990.

Junger, Sebastian. *The Perfect Storm.* New York: Harperperennial, 1999.

Kelly, Orr. *From a Dark Sky: The Story of the US Air Force Special Operations.* Novato, CA: Presidio Press, 1996.

Keogan, Joseph. *The Igor I. Sikorsky Aircraft Legacy: The Chronology of Fixed-Winged and Rotary-Wing Aircraft of Igor I. Sikorsky and the Sikorsky Aircraft Company.* Stratford, CT: Igor I. Sikorsky Historical Archives, 2003.

Kuznetsov, G. I. *Kamov OKB 50 Years, 1948–1998.* Edinburgh, Scotland: Polygon Publishing House, Birlinn Publishing, 1999.

Landis, Tony, and Dennis Jenkins. *Lockheed AH-56A Cheyenne.* North Branch, MN: Specialty Press, 2000.

Liberatore, E. K. *Helicopters before Helicopters.* Malabar, FL: Krieger Publishing Company, 1998.

Lightbody, Andy, and Joe Poyer. *The Illustrated History of Helicopters.* Lincolnwood, IL: Publications International, 1990.

Marshall, Chris. *The Defenders.* London: Aerospace Publishing, 1988.

Mashman, Joe (as told to R. Randall Padfield). *To Fly Like a Bird.* Potomac, MD: Phillips Publishing, 1992.

McGuire, Francis G. *Helicopters 1948–1998: A Contemporary History.* Alexandria, VA: Helicopter Association International, 1998.

Mesko, Jim. *Airmobile: The Helicopter War in Vietnam.* Carrollton, TX: Squadron Signal Publications, 1984.

Novosel, Michael J. *Dustoff: The Memoirs of an Army Aviator.* Novato, CA: Presidio Press, 1999.

O'Grady, Scott. *Return with Honor.* New York: Harper, 1996.

Oren, Michael B. *Six Days of War: June 1967 and the Making of the Modern Middle East.* New York: Ballantine Books, 2003.

Palmer, Norman, and Floyd D. Kennedy, Jr. *Military Helicopters of the World.* Annapolis, MD: Naval Institute Press, 1984.

Pearcy, Arthur. *U.S. Coast Guard Aircraft since 1916.* Shrewsbury, England: Airlife Publications, 1991.

Pember, Harry. *Seventy-five Years of Aviation Firsts.* Stratford, CT: Sikorsky Historical Archives, 1998.

Ripley, Tim. *Jane's Pocket Guide: Modern Military Helicopters.* London: Jane's, 1997.

Rogers, Mike. *VTOL Military Research Aircraft.* Somerset, England: Haynes and Co., 1989.

Ross, Frank, Jr. *Flying Windmills.* New York: Lothrop, Lee, & Shepard Co., 1953.

Saunders, George H. *Dynamics of Helicopter Flight.* New York: John Wiley and Sons, 1975.

Shrader, Charles R. *The First Helicopter War: Logistics and Mobility in Algeria, 1954–1962.* Westport, CT: Greenwood Publishing, 1999.

Sikorsky, Igor I. *The Story of the Winged-S.* Rev. ed. New York: Dodd, Mead & Company, 1967.

Simpson, R. W. *Airlife's Helicopters & Rotorcraft: A Directory of World Manufacturers and Their Aircraft.* Shrewsbury, England: Airlife Publishing Ltd., 1998.

Smith, J. Richard. *Focke-Wulf: An Aircraft Album No. 7.* New York: Arco Publishing Co., 1973.

Smith, J. R., and Antony Kay. *German Aircraft of the Second World War:* London: Nautical & Aviation Publishing Company, 1972.

Spenser, Jay P. *Vertical Challenge: The Hiller Aircraft Story.* Seattle: University of Washington Press, 1992.

Stanton, Shelby L. *Anatomy of a Division: The First Cav in Vietnam.* Novato, CA: Presidio Press, 1987.

Stapfer, Hans-Heiri. *Mi–24 Hind in Action.* Carrollton, TX: Squadron/Signal Publications, 1988.

Swanborough, Gordon, and Peter M. Bowers. *United States Navy Aircraft since 1911.* London: Putnam, 1990.

Taylor, Michael J. H., ed. *Jane's Encyclopedia of Aviation.* New York: Portland House, 1989.

Tilford, Earl H., Jr. *The USAF Search and Rescue in Southeast Asia.* Washington, DC: Office of Air Force History, 1992.

Townson, George. *Autogiro: The Story of the Windmill Plane.* Fallbrook, CA: Aero Publishers, 1985.

Uttley, Matthew R. *Westland and the British Helicopter Industry, 1945–1960: Licensed Production vs. Indigenous Innovation.* London: Taylor and Francis, 2001.

Weinert, Richard P., Jr., and Susan Canedy, eds. *A History of Army Aviation—1950–1962.* Ft. Monroe, VA: Office of the Command Historian, United States Army Training and Doctrine Command, 1991.

Wood, Derrick. *Jane's World Aircraft Recognition Handbook.* 4th ed. Colsdon, Surrey, UK: Jane's Information Group, 1989.

Young, Ralph B. *Army Aviation in Vietnam, 1963–1966: An Illustrated History of Unit Insignia Aircraft Camouflage and Markings.* 2 vols. New Jersey: Huey Company, 2000.

Young, Warren R., et al. *The Epic of Flight: The Helicopters.* Alexandria, VA: Time-Life Books, 1982.

JOURNALS AND MAGAZINES

Apostolo, Giorgio, and Robert A. Hasskarl, Jr. "Early Military Use of Rotary-Wing Aircraft." *Air Power Historian* 12, no. 3 (July 1965): 75–77.

Carlson, Ted. "Marine Twin Hueys." *World Airpower Journal* 42 (autumn/fall 2000).

Glines, C. V. "The Skyhook." *Air Force Magazine* 71 (July 1988).

Gordon, Yefim, and Dimitriy Komissarov. "Mil Mi-24 'Hind.'" *World Airpower Journal* 37 (summer 1999).

Halcomb, Mal. "The Development History of the Helicopter in Germany from WWI to the End of WWII." *Airpower* 37 (March 1990).

Hasskarl, Robert A. Jr. "Early Military Use of Rotary-Wing Aircraft." *The Air Power Historian.* 12, no. 3, July 1965.

Naylor, Sean D. "In Shah-E-Kot, Apaches Save the Day—and Their Reputation." *Army Times,* March 25, 2002.

"The Wing Slip." Aviator's Post #743, American Legion, Massapequa, NY. #576, May 18, 1984.

WEBSITES

Bell Helicopter-Textron, http://www.bellhelicopter.textron.com.

The Boeing Company, http://www.boeing.com.

Brain, Marshall. "How Helicopters Work," http://www.howstuffworks.com/helicopter.htm.

British Army Air Corps Association, http://www.aacn.org.uk.

British Army Air Corps Historical Flight, http://www.rdg.ac.uk

British Army Air Corps Museum, http://www.flying-museum.org.uk.

Fort Rucker, AL. Home of U.S. Army Aviation, www-rucker.army.mil.

Helicopter Museum, http://www.hmfriends.org.uk.

Helicopter World, helicopter.virtualave.net.

"Helicopters of the U.S. Army," http://www.geocities.com/capecanaveral/hangar/3393/army.html.

Helicopter's History Site, http://www.helis.com.

Igor Sikorsky Historical Archives, Inc., http://www.iconn.net/igor/index lnk.html.

International Helicopters, http://www.globalsecurity.org.

Leishman, J. Gordon. "Evolution of Helicopter Flight," http://www.flight 100.org/history/helicopter.html.

"Pitcairn PCA-1A," National Air and Space Museum, http://www.nasm.edu/nasm/aircraaft/pitcairn_pca.htm.

Royal Air Force Official Site, http://www.raf.mod.uk.

Royal Navy Official Site, http://www.royal-navy.mod.uk.

Russian Aviation Museum, www2.ctrl-c.liu.se/misc/ram.

Sikorsky Aircraft Corporation, http://www.sikorsky.com.

Soviet/Russian Helicopters, http://www.royfc.com/links/acft_coll.

United Kingdom, British Army Official Site, http://www.army.mod.uk.

Unusual Aircraft, http://www.unrealaircraft.com.

U.S. Air Force Historical Research Agency, http://www.au.af.mil/au/afhra/.

U.S. Air Force Museum, http://www.asc.wpafb.af.mil/museum.

U.S. Army Aviation and Missile Command (AMCOM), http://www.redstone.army.mil/history/aviation/.

INDEX

ABOUT THE AUTHOR

DR. STANLEY S. McGOWEN is adjunct professor of military history at Columbia College, Joint Reserve Base, Fort Worth, Texas. Dr. McGowen is a twenty-one year veteran of the U. S. Army, with various assignments in the United States, Vietnam, Germany, and Alaska. In addition to serving in the infantry and military intelligence, he is an Army aviator with over 5,000 hours of flying time in helicopters. His published works include *Horse Sweat and Powder Smoke: The First Texas Cavalry in the Civil War* (1999), which won the Summerfield G. Roberts Writing Award for the best book about Texas and Texans in the Civil War. He has also published numerous articles in professional journals, as well as contributing to encyclopedias of World War II, Korea, and Vietnam